ʿAlī ibn Sahl Rabban aṭ-Ṭabarī's *Health Regimen* or "Book of the Pearl"

Islamic Philosophy, Theology and Science

TEXTS AND STUDIES

Edited by

Hans Daiber
Anna Akasoy
Emilie Savage-Smith

VOLUME 115

The titles published in this series are listed at *brill.com/ipts*

ʿAlī ibn Sahl Rabban aṭ-Ṭabarī's *Health Regimen* or "Book of the Pearl"

Arabic Text, English Translation, Introduction and Indices

By

Oliver Kahl

TUTA SUB AEGIDE PALLAS · 1683 ·

BRILL

LEIDEN | BOSTON

Cover illustration: *al-Lu'lu'a fī ḥifẓ aṣ-ṣiḥḥa* "(Book of) the Pearl on the preservation of health", text by © Oliver Kahl, 2020; calligraphy by © Nihad Nadam, 2020.

Library of Congress Cataloging-in-Publication Data

Names: Ṭabarī, 'Alī ibn Sahl Rabbān, active 9th century, author. | Kahl, Oliver, translator. | Ṭabarī, 'Alī ibn Sahl Rabbān, active 9th century. Kitāb al-lu'lu'ah.
Title: 'Alī ibn Sahl Rabban aṭ-Ṭabarī's Health regimen or "Book of the pearl" : Arabic text, English translation, introduction and indices / by Oliver Kahl.
Other titles: Kitāb al-lu'lu'ah. English | Health regimen
Description: Boston : Brill, 2021. | Series: Islamic philosophy, theology and science, 0169-8729 ; 115 | Includes bibliographical references and index. | In English and Arabic.
Identifiers: LCCN 2020046146 (print) | LCCN 2020046147 (ebook) | ISBN 9789004445888 (hardback) | ISBN 9789004445895 (ebook)
Subjects: LCSH: Medicine, Arab. | Health–Handbooks, manuals, etc.
Classification: LCC R128.3 .T3313 2021 (print) | LCC R128.3 (ebook) | DDC 610–dc23
LC record available at https://lccn.loc.gov/2020046146
LC ebook record available at https://lccn.loc.gov/2020046147

Typeface for the Latin, Greek, and Cyrillic scripts: "Brill". See and download: brill.com/brill-typeface.

ISSN 0169-8729
ISBN 978-90-04-44588-8 (hardback)
ISBN 978-90-04-44589-5 (e-book)

This book is printed on acid-free paper and produced in a sustainable manner.

Contents

Acknowledgements

This book owes much of its existence to a generous research grant provided by the *Deutsche Forschungsgemeinschaft*. I am indebted to Stefan Weninger, head of the Department of Semitic Studies at Marburg University, for his support, tolerance and trust. Thanks also to Fabian Käs, who readily supplied me with his copy of the manuscript *Ayasofya*, and to Lukas Kahl, who found and scanned some far-off literature. I greatly benefited from being able to use the fine collection of Oriental books held in the University Library of Marburg. I am grateful to the editors of IPTS for including my work, once again, into this series; to the anonymous readers for their comments; and to Abdurraouf Oueslati and Cas Van den Hof for steering the book safely through the various stages of production. If minor disruptions along the way have in retrospect been an incitement rather than an obstacle, this is largely due to Karine, my companion and muse.

O. Kahl

Introduction

الطبيب الجاهل مستحثّ الموت

The ignorant physician is an instigator of death

ʿALĪ IBN SAHL RABBAN AṬ-ṬABARĪ *apud* IAU 2.2/759,6

1 Hygienics in the Arabic Tradition

Before turning in greater detail to its author, its literary structure and, finally, its general epistemological value, the text edited and translated here might benefit from being placed, right at the beginning, into a broad medico-historical frame. ʿAlī ibn Sahl Rabban aṭ-Ṭabarī's (hereafter referred to as Ṭabarī)[1] *Health Regimen* is preserved in the form of two Arabic manuscripts.[2] Ṭabarī wrote it between the years 235/850 and 247/861,[3] after he had completed his chef-d'œuvre *Paradise of Wisdom* and before the death of the caliph al-Mutawakkil, the patron to whom the *Health Regimen* is addressed. The title of the work is given, in one manuscript, as *Fī Ḥifẓ aṣ-ṣiḥḥa* "On the Preservation of Health" and, in the other manuscript, as *Kitāb al-Luʾluʾa* "Book of the Pearl". In it, Ṭabarī treats all aspects of general hygiene, that is preventive health care, according to the medico-philosophical principles and scientific paradigms recognized in his day.[4] As will be discussed later on,[5] Ṭabarī's *Health Regimen* is important for two basic reasons. First, and unlike similar texts of that nature, it is deliberately designed for the profit of the educated (and affluent) layman, rather than the initiated fellow; this design is reflected in the use of largely non-technical language, the presence of a strong didactic undercurrent, the employment of anecdotal set-pieces and references to everyday life, and the overall compositional arrangement of the text. Second, it is the oldest extant (and authentic) Arabic writing dedicated specifically to preventive medicine—

1 The system of transliteration used in this book is that of the *Deutsche Morgenländische Gesellschaft*.

2 For a concise description of these manuscripts see p. 14 below.

3 Dates separated by a slash refer to the Islamic and Christian calendars respectively.

4 For an overview of the work's topical range see the tables on pp. 19 ff. below.

5 See pp. 26 ff. below.

 | DOI:10.1163/9789004445895_002

there are only two potential Arabic precursors of the genre proper: a famous treatise on hygiene titled *ar-Risāla ad-dahabīya* "The Golden Epistle", which the eighth Shiite imam ʿAlī ibn Mūsā ar-Riḍā (d. 203/818) allegedly composed for the caliph al-Maʾmūn, but which in all likelihood is a pseudepigraph;[6] and Salmawaih ibn Bunān's (d. 225/840) lost *Tadbīr aṣ-ṣiḥḥa* "The Management of Health", whose import and size are completely unknown.[7] Ṭabarī refers to neither of these and, judging from the sources that underlie both his *Paradise of Wisdom* and his *Health Regimen*,[8] he also seems to have disregarded the cluster of medico-religious traditions (*aḥādīṯ*) which were later first isolated and then reassembled to form the illustrious genre of 'prophetic medicine' (*aṭ-ṭibb an-nabawī* or *ṭibb an-nabīy*), itself an exponent of theology rather than medicine. It is clear that hygienics, as it would have been understood in Ṭabarī's time, involves virtually all branches of preventive medicine: dietary, sexual, somnial, seasonal and daily regimen; basic humoral physiology and pathology; phlebotomy and catharsis; physiognomy and psychosomatics; uroscopy; social and environmental factors; organotherapy and sympathetic prescriptions; and the application of both simple and compound drugs. In this general sense, the topical scope of Ṭabarī's *Health Regimen* is just as wide as the spectrum of sources he employed, be it explicitly or implicitly; in a narrower sense, however, and focusing on potential textual templates in which these various branches of preventive medicine appear *in already synthesized form*, we can identify only a small number of concrete texts, all of which are either of Greek or Indian origin and were, in principle, available to Ṭabarī through Syriac or Arabic transla-

6 For Arabic manuscripts of this text see UllMed 190 note 2, for an early Persian translation see FoPM 50 no. 69; cf. also GaS 3/226 and, for a recent study, SpRis 8 and *passim*.

7 Registered GaS 3/227 no. 2. It should further be noted that Qusṭā ibn Lūqā's (fl. c. 205/820–300/912) *Kitāb fī Ḥifẓ aṣ-ṣiḥḥa wa-izālat al-maraḍ* "Book on the Preservation of Health and the Elimination of Illness" (UllMed 190 and GaS 3/271 no. 9), as well as Isḥāq ibn ʿImrān's (d. before 296/909) *Risāla fī Ḥifẓ aṣ-ṣiḥḥa* "Treatise on the Preservation of Health" (UllMed 190), are possible precursors only in the tightest chronological terms. Another treatise on hygiene that is preserved in Persian s.t. *Risāla-ye ḥifẓ aṣ-ṣiḥḥa* (var. *Kitāb-e iḫtiyārāt-e taqwīm*), and which Adolf Fonahn (FoPM 52 no. 77) fleetingly considered to be the translation of a writing by Ḥunain ibn Isḥāq (d. 260/873), must be a later production, as it contains several references to Rhazes, who was a child when Ḥunain died (cf. UllMed 190). Finally, the famous epistle (*risāla*) which Masīḥ ad-Dimašqī (d. after 225/840) wrote, a few decades earlier, for the caliph Hārūn ar-Rašīd (reg. 170/786–193/809), and which Ṭabarī, on one occasion (§§73–74), implicitly used as a source, is not only a much larger composition but also, and more importantly, conceived as a *general* medical handbook, rather than a synopsis focusing on preventive medicine; for this work see bibliography s.n. DimRH.

8 For Ṭabarī's sources see pp. 14–17 below.

tions: Hippocrates' *Περὶ διαίτης ὑγιεινῆς* (Arabic title *Ḥifẓ aṣ-ṣiḥḥa*)[9] and *Ὑγιεινόν* (Arabic title *Ḥifẓ aṣ-ṣiḥḥa*),[10] Rufus of Ephesos' *Ὑγιεινά* (Arabic titles *Ḥifẓ aṣ-ṣiḥḥa* or *Waṣāyā ḥifẓ aṣ-ṣiḥḥa* or *Kunnāš Siyāsat aṣ-ṣiḥḥa*),[11] and Galen's *Περὶ τῶν ὑγιεινῶν πραγματεία* (Arabic titles *Tadbīr al-aṣiḥḥā'* or *al-Ḥīla li-ḥifẓ aṣ-ṣiḥḥa*);[12] Ayurvedic templates[13] include the relevant sections on hygienics (*svasthavṛtta*) in Suśruta's *Saṃhitā*,[14] Caraka's *Saṃhitā*,[15] notably Vāgbhaṭa's *Aṣṭāṅgahṛdayasaṃhitā*,[16] and to a lesser extent Mādhava's *Nidāna*.[17]

On the background of the foregoing observations, the importance of Ṭabarī's *Health Regimen* is almost self-evident; and yet it remains to say a few words of justification with regard to our current edition and translation of this text. A mere glance at the comparative chart in the appendix to the present book[18] will immediately reveal that Ṭabarī's *Health Regimen*, on the level of informational content, is a redesigned abstract of his much larger *Paradise of Wisdom*,[19] and that it offers new material only to a very limited extent; at the same time, the *Health Regimen* is already available in a German translation based directly on the two extant Arabic manuscripts.[20] The legitimate question arises as to whether an Arabic edition and English translation of the *Health Regimen* is sufficiently valuable at all. In response I should like to remark the following: first, the existing Arabic edition of the *Paradise of Wisdom* suffers from such an amount of philological problems that the text can only be used with the greatest circumspection and, at times, be hardly understood—this is why Carl Brockelmann, in his review of this book and barely four years after its publication, already called for a new, 'definitive' edition;[21] second, and more importantly, the *Health Regimen* has never before been edited in its original Arabic form, and the existing German translation of it is neither always reliable nor

9 FiCH 41 no. 26.
10 FiCH 81 no. 162 (lost in Greek).
11 UllMed 74 no. 15.
12 FiCG 37 no. 37.
13 Explicitly cited by Ṭabarī in the *Paradise of Wisdom*, see ṬabFir 557,11f. and *passim*.
14 SuSaṃ 1/36–45 and 370–457.
15 CaSaṃ 1/105–162.
16 VāgAṣṭ 1/22–52.
17 MāNid *passim*. For the transmission of these Sanskrit works into Arabic see RhaCB 7–27.
18 See pp. 208–214; cf. also pp. 21ff. for a more thorough discussion of these findings.
19 Edited in 1928 by Muḥammad Zubair aṣ-Ṣiddīqī, cf. bibliography s.n. ṬabFir.
20 Translated in 1975 by Usāma Raslān, cf. bibliography s.n. ṬabHT.
21 "Wenn diese [Ausgabe] auch genügt, um über den literarischen Charakter des Werkes vorläufig zu orientieren, so kann sie doch eine endgültige Ausgabe nicht ersetzen, für die eine neue Vergleichung der benutzten Hdss. wahrscheinlich schon genug Material liefern würde" (BroFḤ 288); cf. also the judgements SilB 3,-4f. and SchṬab 4,13f. About 20% of

furnished with an adequate set of explanatory notes. Yet even aside from these considerations, Ṭabarī's *Health Regimen* is well worth to be studied in its own right—as the earliest surviving Arabic specimen of the genre, and as a rare example of a medical text conceived explicitly with a view to non-professional circles.

2 ʿAlī ibn Sahl Rabban aṭ-Ṭabarī

a *Life*

Information about Ṭabarī's life can be gleaned from a variety of medieval Arabic and Persian sources, but the overall picture that emerges from these terse and repetitive accounts remains extremely scrappy and, with regard to certain biographical details, downright inconsistent; random *autobiographical* remarks, scattered loosely throughout Ṭabarī's extant writings, may on occasion provide a few missing pieces to the puzzle. This is, to be sure, by no means unusual for a scholar of Ṭabarī's time and place, even though that lack of sufficiently reliable biographical data is strangely contrasted by the high-ranking positions he held and by the undisputed importance of his medical and philosophical legacy. Ever since the coincidental publication of two of Ṭabarī's main works in the 1920s,[22] several modern researchers have now and then attempted, with more or less diligence and erudition, to give a reasonably coherent and comprehensive narrative of his life—here, I would like to single out the efforts of Alphonse Mingana,[23] Max Meyerhof,[24] Muḥammad Zubair aṣ-Ṣiddīqī[25] and, most notably and most recently, those of Rifaat Ebied and David Thomas.[26] As the last-named biographical essay by Ebied and Thomas is very thorough and up to date, I will confine myself in the following to a more concise and conflated sketch of Ṭabarī's life, departing always from the indigenous sources.[27]

---

the *Paradise of Wisdom* have been translated over the years, by different scholars, mostly into German, see UllMed 120f. note 4 with 346 (addenda) and GaS 3/239; further SiPK 3. A new but even less critical edition of the *Paradise of Wisdom* came out some years ago in Beirut, cf. bibliography s.n. ṬabFir²; this presumably well-meaning but irredeemably flawed publication is not cited in the present book.

22 The *Kitāb ad-Dīn wad-daula* "Book of Religion and Empire", edited and translated in 1922–1923, cf. bibliography s.n. ṬabDD; and the *Firdaus al-ḥikma fī ṭ-ṭibb* "Paradise of Wisdom about Medicine", edited in 1928, cf. bibliography s.n. ṬabFir.

23 ṬabDD x–xvi.

24 MeyṬab 43–56.

25 SidML 46–52; also ṬabFir (editor's preface) pp. د–ط.

26 EbThPW 2–24. Cf. further UllMed 119f., GaS 3/236f. and ThoṬab *passim*.

27 These are (in chronological order): Abū Ǧaʿfar aṭ-Ṭabarī (d. 310/923) *Taʾrīḫ ar-rusul wal-*

Right at the beginning, however, we need to dispose of two misconceptions. First, the Arab writer Ǧamāladdīn al-Qifṭī (d. 646/1248), referring to Muḥammad ibn Isḥāq an-Nadīm's (d. 380/990) 'book' (*kitāb*), explains in a passage which is not actually found in the existing edition(s) of the *Fihrist* that Ṭabarī's father Sahl was called *Rabban* because he was a Jewish Rabbi (*rabbīn al-yahūd*);[28] this prompted Moritz Steinschneider, and others in his wake, to declare Ṭabarī the son of a "Rabbiner" and, by extension, a Jew himself.[29] Carl Brockelmann and Max Meyerhof have shown, already in the early 1930s, that this claim is false[30]—it may originate, insofar as al-Qifṭī is concerned, in an old interpolation at the beginning of Ṭabarī's own *Paradise of Wisdom*, according to which his father was well acquainted with, inter alia, the Hebrew language (*al-ʿibrānīya*).[31] *Rabban* is, in fact, a honorific Christian title, representing an Arabic transliteration of Syriac ܪܒܢ "our master"[32] (not רב[נו]). This is confirmed by other Arabic sources which explicitly call Ṭabarī "a Christian";[33] by certain turns of phrase which Ṭabarī himself employs in the *Paradise of Wisdom*, completed before his conversion to Islam;[34] and by the fact that in this work he amply quotes from the Bible (not from the Koran and certainly not from the Torah). Lastly, we possess two apologetic-polemical writings which Ṭabarī composed after his conversion to Islam, and which are dedicated to an examination of contradictions in Christian theology[35]—works of that nature are unlikely to

mulūk [Ṭab^{2}Ann]; an-Nadīm (d. 380/990) *al-Fihrist* [NadFih]; al-Baihaqī (d. 565/1170) *Tatimma-ye Ṣiwān al-ḥikma* [BaiTat]; Ibn Isfandiyār (d. after 613/1216) *Taʾrīḫ-e Ṭabaristān* [IsHṬ]; Yāqūt al-Ḥamawī (d. 626/1229) *Iršād al-arīb* [YāIr]; al-Qifṭī (d. 646/1248) *Taʾrīḫ al-ḥukamāʾ* [ZauMuḫ]; Ibn Abī Uṣaibiʿa (d. 668/1270) *ʿUyūn al-anbāʾ* [IAU]; Ibn Ḫallikān (d. 681/1282) *Wafayāt al-aʿyān* [ḪaWaf].

28 ZauMuḫ 231,17 ff.

29 StALJ 32 no. 20 (published in 1902).

30 BroFḤ 270 and, notably, MeyṬab 43.

31 ṬabFir 1 note ١ (rejected but documented by the editor of the Arabic text).

32 See e.g. PSThes 2/3788: "*magister noster* [...] vox *Raban* e consuetudine Syrorum presbyteris et monachis attribui solet"; further MeyṬab 44,15–24. Ṭabarī himself says that the word *Rabban* means "our great (master) and our teacher" (*ʿaẓīmunā wa-muʿallimunā*), see ṬabFir 1,15.

33 Ṭab^{2}Ann 3.2/1283,19 and 1293,9 (*naṣrānī*); NadFih 1/316,20 (*naṣrānī*); ḪaWaf 4/245,10 (*masīḥī*). Perhaps al-Baihaqī's (d. 565/1170) somewhat odd statement that Ṭabarī knew "tricks" (الحيل) should indeed be emended to read "(the) Gospel" (الانجيل), as the editor of the Persian text suggests, see BaiTat 14 no. 6,3 (note that al-Baihaqī's short biographical account, according to Ebied and Thomas [EbThPW 3], "evidently [?] confuses" information about Ṭabarī and his father Sahl).

34 For example ṬabFir 555,4 (referring to God): *al-ḫallāq al-ḥakīm tabāraka smuhū wa-taʿālā ḏikruhū* "the wise Creator—blessed is His name, sublime His invocation".

35 Cf. p. 10 f. nos. 2 and 12.

have been written by a converted Jew.[36] The second misconception is chronological: some Arabic sources maintain that the famous physician, alchemist and philosopher Abū Bakr Muḥammad ibn Zakarīyāʾ ar-Rāzī (Rhazes, d. 313/925) was a student of Ṭabarī, especially in the field of medicine.[37] Rhazes was born in Rayy in the year 251/865;[38] and whilst we have no certainty about the date of Ṭabarī's demise, there is, as we shall see further on, a general consensus that he cannot possibly have outlived the year 256/870—this means that Rhazes was still a small child at the lattermost time of Ṭabarī's death (if he had not died already). The alleged teacher-pupil relationship between Ṭabarī and Rhazes must therefore be interpreted, not as a historical fact but rather as a token of intellectual influence.[39]

Even a modest reconstruction of Ṭabarī's life is full of yawning gaps. His complete name is Abū l-Ḥasan ʿAlī ibn Sahl Rabban aṭ-Ṭabarī, and he was "born and raised" in Tabaristan,[40] an old Iranian region bordering on the southern shores of the Caspian Sea. The date of his birth is not recorded in the sources, but the year 174/790, or thereabout, seems to be the most informed (and most recent) guess.[41] His father Sahl was a philosophically-minded Syrian Christian scholar, probably a Nestorian, who had moved to Tabaristan from the Central Asian city of Merv;[42] in his new home, he gave up "the trade of his ancestors"

36 As is to be expected, most Arabic and Persian sources struggled with the name *Rabban* (ربن), representing it as *Zyn* (زين) [Ṭab^{2}Ann 3.2/1283 app. *p* and 1293 app. *k*; BaiTat 14 no. 6,1; IsHṬ 80 no. 1,1; YāIr 6/429,13 and 460,19; ZauMuḫ 231 app. *a* and *c*; ḤaWaf 4/245,9], *Rzyn* (رزين) [Ṭab^{2}Ann 3.2/1283 app. *p* and 1293 app. *k*], *Zyl* (زيل) [NadFih 1/296,10], *Rbl* (ربل) [NadFih *apud* IAU 2.2/759,2], *Zyd* (زيد) [ZauMuḫ 231 app. *a*], *Ryn* (رين) [NadFih 1/316,20], or *Zbl* (زبل), *Dbl* (دبل) and *D·l* (دبل) [NadFih 2/141]. The two manuscripts of the *Health Regimen*, too, offer *Rayan* (ريّن Oxford, sic) and, respectively, *R·n* (رن Istanbul, sic).

37 ZauMuḫ 231,10 f.: *ḫarağa* […] *ilā r-Raiy fa-qaraʾa ʿalaihī Muḥammad ibn Zakarīyāʾ ar-Rāzī wa-stafāda minhū ʿilm^{an} kaṯīr^{an}* "he (Ṭabarī) emigrated to Rayy, where Muḥammad ibn Zakarīyāʾ ar-Rāzī read under him, thus acquiring a great deal of knowledge"; IAU 2.2/759,4: *wa-huwa muʿallim ar-Rāzī ṣināʿat^{a} ṭ-ṭibb* "and he (Ṭabarī) was the teacher of ar-Rāzī in the art of medicine"; ḤaWaf 4/245,9 (in the biography of Rhazes): *wa-kāna štiġāluhū biṭ-ṭibb ʿalā* […] *aṭ-Ṭabarī* "and he (Rhazes) studied medicine under aṭ-Ṭabarī".

38 BīFih 4,8 f.

39 So already pointed out by Fuat Sezgin, see GaS 3/237,21 ff.; similarly ThoṬab 18a.

40 IAU 2.2/759,4 f.; cf. also ZauMuḫ 231,9.

41 EbThPW 13; previous estimates range from 158/775 to 194/810 (e.g. ṬabFir p. و [revised SidML 48], MeyṬab 45 and 47, UllMed 119, GaS 3/237).

42 Meyerhof is inclined to believe that Ṭabarī was born in Merv and that he came to Tabaristan as a child, in the company of his father (MeyṬab 45 f.). This assumption is based on an ambiguous account about a celestial event in the skies over Merv, datable to the year 202/818, which Ṭabarī himself recorded in the *Paradise of Wisdom* (ṬabFir 519,22–520,1) and which Meyerhof takes as a firsthand observation made by either father or son;

(*ṣināʿat ābāʾihī*) in order to practise medicine[43]—it was largely to this activity, which he pursued with skill and altruism, that he owed the title *Rabban*;[44] he was known moreover as an astronomer.[45] Ṭabarī's paternal uncle, too, a man called Abū Zakkār Yaḥyā ibn Nuʿmān, was a scholar, renowned apparently, from Khorasan (viz. Merv) to Iraq, for his superior intelligence and rhetorical abilities, as well as for being the author of a (lost) theological writing.[46] Ṭabarī speaks very highly, and affectionately, of his father who, so he says, "taught me, from when I was little, as much about (his science) as was possible in view of the talents given to me by God and in accordance with what time and nature permitted";[47] it is therefore no surprise to find prescriptions in the *Paradise of Wisdom* that Ṭabarī explicitly, and proudly, linked to his father.[48] Like most Irano-Syrian Christians in his day, Ṭabarī had two native languages, Persian and Syriac,[49] but he must have learned Arabic already at an early age, considering his mastery of that language as displayed in his extant writings;[50] he does, on the other hand, not seem to have had direct access to Greek, nor is it obvious that he was acquainted with Sanskrit.[51] We hear nothing further about Ṭabarī's youth and early adulthood. It is possible that he, like his father,

---

straight after, however, Ṭabarī clarifies that he obtained this and similar accounts "be it from eyewitnessing or be it from hearsaying, which (latter) may substitute eyewitnessing [!]" (*immā ʿiyān wa-immā samāʿ yaqūm maqām al-ʿiyān*), see ṬabFir 520,1f. Meyerhof's proposition is obviously linked to his much later estimate for Ṭabarī's birth (c. 194/810).

43 ṬabFir 1,14; the 'ancestral trade' was scribehood, see ṬabFir 1,11f. with BaiTat 14 no. 6,2.

44 ṬabFir 1,15.

45 SutMA 14 no. 25. The famous astrologer Abū Maʿšar al-Balḫī (d. 272/886) apparently even knew a (Syriac?) version of Ptolemy's *Almagest* made by Sahl, see PinAM 17 with note 3 and 56 with note 4.

46 ṬabDD 1/147 and 152 (English translation) = 2/124 and 129 (Arabic text).

47 ṬabFir 1,16f.; al-Qifṭī (d. 646/1248), having introduced Ṭabarī as an eminent physician, says that he also "read philosophy and excelled at the natural sciences" (*yaqraʾ ʿilm al-ḥikma wa-nfarada biṭ-ṭabīʿīyāt*), see ZauMuḫ 231,9f.

48 For example ṬabFir 488,6–16.

49 His transliterations of Persian names and terms—and their relative preponderance—in the *Paradise of Wisdom* are remarkably accurate; besides, Persian would have been the language of daily commerce in his native land. His command of Syriac, probably spoken at home and in Christian circles, is also evidenced by the fact that he himself translated the *Paradise of Wisdom* from Arabic into Syriac (cf. p. 10 no. 3b), as well as by references to Syriac compendia (*kunnāšāt* [...] *li-ahl sūriyā*) which he confirms to have read (ṬabFir 1,19).

50 The anecdote recorded by Ibn Isfandiyār (d. after 613/1216), according to which Ṭabarī excused the semantic poverty of his written Arabic by the fact that it was not his own language (IsHṬ 43), is contaminated (cf. MeyṬab 53f. and EbThPW 3); elsewhere, the same Ibn Isfandiyār praises Ṭabarī's originality and eloquence (IsHṬ 80).

51 On the question of Ṭabarī's linguistic abilities see also the perceptive remarks in SchṬab

practised medicine for a while, but he must at some point, probably in his thirties and for reasons unknown, have entered the service of local rulers[52] in an administrative capacity, or else he would not have risen to be secretary (*kātib*) of the governor (*spāhbed*) of Tabaristan, Māzyār ibn Qārin, and a trusted member of the latter's inner circle.[53] In 225/840—the year in which the rebellious Māzyār was overthrown, deported to Iraq and flogged to death—Ṭabarī fled to Rayy and, soon afterwards, on to Samarra, the recently founded new capital of the ʿAbbāsid empire.[54] According to a credible scribal note on the first manuscript leaf of Ṭabarī's *Book of Religion and Empire*,[55] he was pardoned and then given employment by the caliph al-Muʿtaṣim (d. 227/842);[56] according to one source, this employment initially took the form of a secretarial position in the 'Foreign Office' (*dīwān-e inšāʾ*).[57] An allegation made in an otherwise trustworthy source—namely that Ṭabarī converted to Islam already under al-Muʿtaṣim "who promoted him and favoured him with his presence"[58]—has been shown to be wrong.[59] Ṭabarī stayed in Samarra.[60] We know from a firsthand observation of a comet, which Ṭabarī himself eyewitnessed and recorded in the *Paradise of Wisdom*, that he was in Samarra throughout the reign of al-Muʿtaṣim's successor al-Wāṯiq (reg. 227/842–232/847);[61] and from strong circumstantial evidence we also know that he converted to Islam soon after 235/850, under the caliph al-Mutawakkil (reg. 232/847–247/861) who moreover "admitted him into the circle of his boon companions (*nudamāʾ*)"[62]—at that point, Ṭabarī would have been in his early sixties.[63] Considering the nature of

---

12–20 and 47 ff. (Schmucker does not rule out that Ṭabarī knew at least some Sanskrit, ibid. 41).

52 ZauMuḫ 231,9: *yataṣarraf fī ḫidmat wulāt* [*Ṭabaristān*].

53 Ṭab²Ann 3.2/1283,19–1284,1 and 1293,9 (sub anno 224/839); NadFih 1/296,11; IsHṬ 80 no. 1; ZauMuḫ 231,19 (after an-Nadīm); IAU 2.2/759,3 (after an-Nadīm).

54 ZauMuḫ 231,10 ff.; this source also links Ṭabarī's flight to Rayy explicitly with Māzyār's revolt (*fitna*), which was crushed by the imperial armies.

55 Cf. p. 10 no. 2.

56 ṬabDD xiv.

57 IsHṬ 43 and 80 (the translation 'Foreign Office' is E.G. Browne's).

58 NadFih 1/296,11 f.; ZauMuḫ 231,19 f. (after an-Nadīm); IAU 2.2/759,3 (after an-Nadīm). Ibn Ḫallikān (d. 681/1282), the only other source that mentions Ṭabarī's conversion to Islam, simply says "he was a Christian, then he became a Muslim", see ḪaWaf 4/245,10.

59 MeyṬab 52 f., ThoṬab 17b, EbThPW 11 ff.

60 ZauMuḫ 231,12.

61 ṬabFir 519,18–22; cf. MeyṬab 54 f.

62 NadFih 1/296,12; ZauMuḫ 231,20 (after an-Nadīm); IAU 2.2/759,3 f. (after an-Nadīm). On the possible reasons for Ṭabarī's late (and voluntary) conversion to Islam see the sagacious and stimulating expositions of Ebied and Thomas (EbThPW 17–23).

63 The problem of dating Ṭabarī's conversion, which is intrinsically connected with the prob-

Ṭabarī's literary output[64] and the esteem in which he seems to have been held at the centre of ʿAbbāsid power, it is likely that he took up medicine again, perhaps even as part of the illustrious group of court physicians.[65] The fact remains that, from here on, the indigenous sources fall silent, and we therefore have no direct information whatsoever about the further course of Ṭabarī's life. A very plausible assumption puts the time of Ṭabarī's death in the decade between 246/860 and 256/870, at which point he would have been somewhere in his seventies;[66] that he died in Samarra is likely.

For a scholar of the 3rd/9th century, Iraq was the place to be. Ṭabarī had no shortage of eminent colleagues, nor a lack of intellectual inspiration; yet he does not seem to have associated himself particularly with any of the various literary and scientific circles which existed in Baghdad and Samarra at the time. Rifaat Ebied and David Thomas have already given some indications as to whom Ṭabarī could and probably would have known,[67] so suffice it here to single out two famous *medical* fellows whose acquaintance Ṭabarī surely would have made, and whom he also quotes in his *Paradise of Wisdom*:[68] Yūḥannā ibn Māsawaih (d. 243/857) and Ḥunain ibn Isḥāq (d. 260/873), both high-ranking physicians to the court;[69] besides, Ṭabarī must have personally known the director (*raʾīs*) of the hospital of Gondēšāpūr, on whose authority he transmits a piece of *oral* information.[70] Finally, whilst Ṭabarī's alleged relationship with Rhazes has been shown to be legendary,[71] it is on the contrary entirely conceivable that he was, for some time, the teacher of his later celebrated com-

lem of determining his lifespan, has been discussed in great detail and brought to a convincing solution by Ebied and Thomas (EbThPW 11 ff.).

64 Cf. pp. 10 ff. below.

65 In one old source, though, Ṭabarī is explicitly placed in the *adab* category, see NadFih 1/296,12 (repeated IAU 2.2/759,4).

66 EbThPW 13; previous estimates range from soon/long after 240/855 to about/not long after 250/864 (SidML 52, UllMed 120, GaS 3/237, ThoṬab 18a).

67 EbThPW 14 f.

68 Cf. p. 16 f. below.

69 In the introduction to his edition and translation of Ḥunain ibn Isḥāq's *Ten Treatises on the Eye*, Meyerhof cautiously suggested that the unnamed scholar who, according to Ḥunain's own statement, encouraged him to finish the hitherto uncompleted work, may have been no other than Ṭabarī (MeyTT xxxi f.); Ḥunain, who eventually completed the book sometime after 246/860 (MeyTT xxxix), describes this person as having "attained an eminent position and a very high rank in being promoted chief of the physicians and philosophers" (translation Meyerhof), see for the full passage MeyTT 127 (English translation) = ١٩٤ (Arabic text). It is quite possible that both Ṭabarī and Ḥunain were pupils of Yūḥannā ibn Māsawaih, as suggested by Meyerhof elsewhere (MeyPW 12).

70 ṬabFir 39,10.

71 Cf. p. 6 above.

patriot, the historian and polymath Abū Ǧaʿfar Muḥammad ibn Ǧarīr aṭ-Ṭabarī (fl. 224/839–310/923)—this Ṭabarī sojourned in Baghdad, as a teenager, in the year 241/855 and again between 244/858 and 248/862, then as a tutor to one of the sons of al-Mutawakkil's vizier Ibn Ḫāqān,[72] perhaps even in Samarra; the geographer and biographer Yāqūt al-Ḥamawī (d. 626/1229), in his account of Abū Ǧaʿfar aṭ-Ṭabarī's life, mentions on two occasions that the latter possessed a copy of the *Paradise of Wisdom* which he had written himself from dictation by its author (*samāʿan lahū*).[73]

b *Works*

Including the text edited and translated here, Ṭabarī wrote about fifteen works of varying lengths and subject matter; most of these works are now lost, and of the four extant ones three have already been published. The following, alphabetically arranged inventory takes all his attested writings into account:

1. *al-Ādāb wal-amṯāl ʿalā maḏāhib al-furs war-rūm wal-ʿarab* "Belles-Lettres and Proverbs in the Traditions of the Persians, Greeks and Arabs" (lost): NadFih 1/316,20 f.; GaS 3/240 no. 10.
2. *ad-Dīn wad-daula* "Religion and Empire" (written c. 241/855): UllMed 120, GaS 3/240 no. 8; **ed.** and **tr.** see bibliography s.nn. ṬabDD and EbThPW².

3a. *Firdaus al-ḥikma* "Paradise of Wisdom" (completed in 235/850 [see ṬabFir 2,22 ff.]): NadFih 1/296,13 f., ZauMuḫ 231,12 ff., IAU 2.2/759 no. 1; UllMed 120 ff., GaS 3/239 no. 1; **ed.** see bibliography s.n. ṬabFir, **tr.** (selections only) see bibliography s.nn. SiGEF, SiIB and SiPK. The title *Baḥr al-fawāʾid* "Ocean of Avails", recorded IsHṬ 80 no. 1, probably refers to the very same work,[74] even though Ibn Isfandiyār lists it *alongside* the "Paradise of Wisdom".

3b. An authorial Syriac translation of the aforesaid work (lost): ṬabFir 8,16.[75]

4. *al-Ǧauhara* "The Gem" (lost; pharmacy?): UllMed 122, GaS 3/239 no. 5.[76]

---

72 See BosṬab 11b.

73 YāIr 6/429,12 ff. (kept under his prayer-carpet) and 460,18 ff. (in six parts).

74 This proposition is based on a remark which Ṭabarī himself makes at the beginning of the *Paradise of Wisdom*, saying that its subtitle (*laqab*) is *Baḥr al-manāfiʿ wa-šams al-ādāb* "Ocean of Benefits and Sun of Refinements", see ṬabFir 8,5 f.; in his Arabic preface, the editor of the *Paradise of Wisdom* already said as much, see ṬabFir pp. ي 19–يا 2.

75 "Then I translated the book [scil. the Arabic original of the *Paradise of Wisdom*] into Syriac" (*ṯumma naqaltu l-kitāb ilā s-suryānīya*).

76 It cannot be completely ruled out that *al-ǧauhara* is a ghost-title, representing an incorrect reading of *al-luʾluʾa* (no. 5 below), viz. اللؤلوه > الجوهره.

5. *Ḥifẓ aṣ-ṣiḥḥa* "The Preservation of Health", variant title *al-Luʾluʾa* "The Pearl" (written between 235/850 and 247/861): IAU 2.2/759 no. 6 (s.n. *Ḥifẓ* ...); UllMed 190, GaS 3/239 nos. 2 and 3; **mss.** Oxford *Bodleian* Marsh 413 and Istanbul *Ayasofya* 3724, **tr.** see bibliography s.n. ṬabHT.
6. *al-Ḥiǧāma* "Cupping" (lost): IAU 2.2/759 no. 8; GaS 3/239 no. 6.[77]
7. *al-Īḍāḥ min as-siman wal-huzāl wa-tahaiyuǧ al-bāh wa-ibṭālihī wa-ǧamīʿ funūnihī* "The Explanation of Fatness and Thinness, and what Excites or Frustrates Sexual Intercourse, and what are all its Kinds" (lost; written before 235/850): ṬabFir 113,9 f.; GaS 3/239 no. 4.
8. *Irfāq al-ḥayāt* "The Value of Life"[78] (lost; medicine or philosophy?): IAU 2.2/759 no. 2; GaS 3/240 (s.n. *ʿIrfān* ...).
9. *Iṯbāt aṭ-ṭibb* "The Confirmation of Medicine" (lost): UllMed 122.
10. *Kunnāš al-ḥaḍra* "The Majesty's Compendium"[79] (lost): NadFih 1/296,14, ZauMuḫ 231,15, IAU 2.2/759 no. 4; GaS 3/240.
11. *Manāfiʿ al-aṭʿima wal-ašriba wal-ʿaqāqīr* "The Useful Properties of Foods, Drinks and Drugs" (lost): NadFih 1/296,14 f., ZauMuḫ 231,15 f., IAU 2.2/759 no. 5; UllMed 264 f., GaS 3/240 no. 7.[80]
12. *ar-Radd ʿalā n-naṣārā* "The Refutation of the Christians" (written after 235/850 and before c. 241/855): UllMed 120, GaS 3/240 no. 9; **ed.** and **tr.** see bibliography s.nn. ṬabRa, ṬabRi and EbThPW[1].
13. *ar-Ruqā* "Magic Spells" (lost): IAU 2.2/759 no. 7; GaS 3/240.

---

77 The unique manuscript of this work which Sezgin, writing in the late 1960s, still registers as extant in Aleppo (after Paul Sbath's [† 1945] *Catalogue*), had vanished without trace in 1971 at the latest, see ṬabHT 12 with note 7.

78 The word *irfāq* "value" could also be read *arfāq*, in which case the title were to be translated as "The Companions of Life".

79 The word *ḥaḍra* "majesty" may also be translated as "(noble) presence"; the term *kunnāš* generally denotes a *medical* compendium. Under different circumstances, the title *Kunnāš al-ḥaḍra*, especially if one accepts an implication of royalty, could be considered a deferential designation of the *Paradise of Wisdom* (no. 3 above), which Ṭabarī wrote at the court in Samarra for the caliph al-Mutawakkil, and which he explicitly calls his *Kunnāš*, see ṬabFir 8,5; however, since all indigenous Arabic sources register the title *Kunnāš al-ḥaḍra* as a separate work, its association with the *Paradise of Wisdom* is rendered highly problematic. Wakelnig's observation that it "seems safe to conclude that *kunnāš* and *Firdaws* refer to one and the same work" (WaṬab 218 note 3) is self-evident but does not necessarily apply to the *Kunnāš al-ḥaḍra*; Olsson opines that the two works are identical (OlDDS 26 with note 61), as did already the editor of the *Paradise of Wisdom* in his Arabic preface (ṬabFir p. ي 3 ff.).

80 The unique manuscript of this work which Sezgin, writing in the late 1960s, still registers as extant in Aleppo (after Paul Sbath's [† 1945] *Catalogue*), had vanished without trace in 1971 at the latest, see ṬabHT 12 with note 7.

14. *Tartīb al-aġḏiya* "The Sequence of Nourishments" (lost): IAU 2.2/759 no. 9; GaS 3/240.
15. *Tuḥfat al-mulūk* "The Gift of the Kings" (lost; topic?): NadFih 1/296,14, ZauMuḫ 231,14 f., IAU 2.2/759 no. 3; GaS 3/240.

It is beyond the scope of the present book to trace the impact of Ṭabarī's medico-philosophical and apologetic-polemical works on subsequent Islamic literature, but it may not be in vain to give here at least some indications into this direction. Generally speaking, it was Ṭabarī's chef-d'œuvre *Paradise of Wisdom*[81] that appears to have attracted the greatest attention on the part of later Arabic- or Persian-writing authors. Rifaat Ebied and David Thomas, talking about Ṭabarī's two apologetic-polemical works *Religion and Empire* and *Refutation of the Christians*,[82] conclude with the following remark: "His two post-conversion writings only began to be noticed (at least in extant works) about a century after his death, and then the *Radd* [scil. *Refutation*] was of interest almost entirely to those who had taken the same steps of conversion; it did not attract a Christian response until a good four centuries after it was written. Similarly, the *Dīn wa-dawla* [scil. *Religion and Empire*] only slowly came to wider notice, and then mainly as a source of proof-texts for Muslims who sought to demonstrate that Muḥammad had been foretold in the Bible".[83] Elvira Wakelnig has, in a recent article and in magnificent detail, investigated the reception of the *Paradise of Wisdom* by later *philosophical* writers and compilers,[84] suggesting that "the *Firdaws* [scil. *Paradise*] was only used as a philosophical source when it was perceived as a purely philosophical text; otherwise, namely when understood as a medical compendium, it was solely quoted for physiological [sic] and medical material".[85]

It remains for us to signal a few non-theological, non-philosophical works in which Ṭabarī's *Paradise of Wisdom* is referred to as a source:[86] here, we can first and foremost direct the reader to the impressive list of largely medico-pharmaceutical writings which Manfred Ullmann registers in his classic *Die Medizin im Islam*—including, inter alia, several dozen quotations from the *Paradise of Wisdom* in ar-Rāzī's (d. 313/925) *al-Ḥāwī*, some in al-Qalānisī's (fl. 590/1194) *al-Aqrabāḏīn*, in Ibn al-Baiṭār's (d. 646/1248) *al-Ǧāmiʿ*, in at-Tīfāšī's

81 Cf. no. 3a in the above inventory.
82 Cf. nos. 2 and 12 in the above inventory.
83 EbThPW 24.
84 WaṬab 222–248.
85 WaṬab 240; similarly ibid. 248,-6 ff.
86 Aside from his *Health Regimen* which is, as far as I can see, quoted nowhere in subsequent literature, the *Paradise of Wisdom* is the only other extant 'medical' writing from Ṭabarī's pen.

(d. 651/1253) *Azhār*, in an-Nuwairī's (d. 733/1333) *Nihāya*, in al-Ğildakī's (d. 743/1342) *Durra*, and in ad-Damīrī's (d. 808/1405) *Ḥayāt al-ḥayawān*.[87] To these we can add an account of an expedition to the summit of Mount Demavend, which the geographer Yāqūt al-Ḥamawī (d. 626/1229) has retained on the authority of Ṭabarī;[88] the same Yāqūt relates, again on the authority of Ṭabarī, a story about a strange bird, native to Tabaristan, called *knkr* (< Persian *kungur* "noctua" [VuLex 2/901a]);[89] and the polymath al-Bīrūnī (d. after 442/1050) relies on Ṭabarī for information about the views of the Indian physician Caraka (fl. c. 50 CE).[90] A systematic inquiry into the literary footprints left by Ṭabarī would surely yield further results.

None of Ṭabarī's works—and here I am thinking particularly of his *Paradise of Wisdom*—were translated into Latin during the medieval or early modern periods. Ferdinand Wüstenfeld and Lucien Leclerc, in 1840 and 1876 respectively, dedicated a few lines to him;[91] but credit is largely due to Edward Granville Browne for having veritably introduced the *Paradise of Wisdom* through one of his lectures held in 1919[92] and, subsequently, for having instigated its edition at the hands of Muḥammad Zubair aṣ-Ṣiddīqī,[93] which came out in 1928—sadly, Browne did not live to see this publication.

3 The *Health Regimen* or "Book of the Pearl"

The four sections that make up this chapter are more specifically intended as prolegomena to the Arabic edition and English translation of Ṭabarī's *Health Regimen*. They cover, in what follows, a brief codicological survey of the two

---

87 UllMed 120 note 4 (the dispensatory of al-Qalānisī and the stone-book of at-Tīfāšī are now edited as well, cf. bibliography s.nn. QalAq and TīfAz).

88 YāBul 2/608,18–609,10 (repeated in al-Qazwīnī's [d. 682/1283] cosmography, see QazKos 1/159,5–17). There is also a report—slightly older than Yāqūt's—about this mountain in Ibn Isfandiyār's (d. after 613/1216) *History of Tabaristan*, who explicitly cites the *Paradise of Wisdom* as his source of information (IsHṬ 35 f.). It seems, however, unlikely that the Arabic versions are directly connected to Ibn Isfandiyār's much shorter Persian passage; besides, Ibn Isfandiyār's claim that the *Paradise of Wisdom* contains this report cannot be verified in the existing edition of the work.

89 YāBul 3/507,3–12. The earlier writer al-Mas'ūdī (d. 345/956), talking about a local bird which he calls *kykm* (?) but which is probably identical with Yāqūt's *knkr* (undotted: كٮكر / كٮكم), also refers to Ṭabarī as a source of information, see MasPO 8/326,2–5.

90 BeInd 1/382 f.

91 WüAN 21 no. 55 and LecHis 1/292 f.

92 BrowAM 37–44.

93 ṬabFir, editor's (unpaginated) preface [p. 1].

manuscript witnesses; an investigation of the text's literary sources; a comprehensive discussion which elaborates on all salient points, describes the compositional structure of the text, sets it against a wider cultural background, and assesses the epistemological value it holds for our understanding of a formative period in Islamic scientific history; the chapter concludes with a short inventory of medicinal weights and measures employed in Ṭabarī's text.

a *The Arabic Manuscripts*

The edition of Ṭabarī's *Health Regimen* is based on the following two sole surviving manuscripts:

> MS Oxford *Bodleian* Marsh 413, fols. 2b,1–31b,13; written by an unnamed copyist in amply dotted, partly vocalized, medium-sized, clear and proficient Nasḫ at 17 lines per page; the title *Fī Ḥifẓ aṣ-ṣiḥḥa* "On the Preservation of Health" appears on fol. 4a,5; the author's name is given on fol. 2b,2 as *Abū ʿAlī ibn Rayan* (sic); no place nor date of copying are recorded (7th/13th century?); the manuscript contains the complete text. For a recent, comprehensive description see SSNC 563ff. (adding UllMed 190 note 5); for a facsimile sample see p. 32 below.

> MS Istanbul *Ayasofya* (now *Süleymaniye*) 3724, fols. 236b,10–246b,15; written by an unnamed copyist in sporadically dotted, largely unvocalized, small, crude and cursory Nasḫ at 21 lines per page; the title *Kitāb al-Luʾluʾa* "Book of the Pearl", together with the author's name *ʿAlī ibn R·n* (رن sic), appear on fol. 236b,10; no place nor date of copying are recorded (8th/14th century?); the manuscript contains the first half of the text (up to and including fol. 16a of the Oxford manuscript). For a very brief description see RiWaSB 835 (whence UllMed 190 note 4 and GaS 3/239 no. 3); for a facsimile sample see p. 33 below.

It is conceivable, though impossible to prove, that the two work titles, as recorded in the manuscript tradition, were originally one, viz. ∗*Kitāb al-Luʾluʾa fī ḥifẓ aṣ-ṣiḥḥa*∗ "Book of the Pearl—on the Preservation of Health".

b *Sources*

As Ṭabarī's *Health Regimen* is, in essence, a redesigned abstract of his *Paradise of Wisdom*, the sources of these two works can be presumed to be essentially the same. In the absence of any systematic inquiries into the textual building blocks of the *Paradise of Wisdom*, and for the limited purposes of our present task, we departed from the short inventory of sources already identified and referenced by Muḥammad Zubair aṣ-Ṣiddīqī, the editor of the *Paradise*, in the (unpaginated) preface to his edition, under the heading "Appendix 2" (see ṬabFir [pp. 4–11], abbreviated below as App2); whichever source could be hitched directly to the *Health Regimen* is further furnished with a reference to the respective paragraph (§) in our own edition/translation of this latter work.[94] Since Ṣiddīqī only referenced authors and texts that are cited frequently and prominently in the *Paradise of Wisdom*,[95] we have provisionally augmented his inventory by incorporating into it (s.n. ṬabFir) some other sources that could be identified thanks to more or less explicit indications in the text. As a general rule, Ṭabarī cites his sources only occasionally, and even then he normally contents himself with naming an author, hardly ever a work title. The following, alphabetical list shows those authors which Ṭabarī explicitly mentions in the *Paradise of Wisdom*, and whose writings implicitly underlie his *Health Regimen*, too. Where traceable, work titles have been added in the form of notes; these notes also contain exemplary references to the *Paradise of Wisdom*, and full references to the *Health Regimen*. A source-critical analysis of the *Paradise of Wisdom* remains a desideratum, and our list does not claim to be exhaustive.

Authorities named by Ṭabarī include: Alexander of Aphrodisias,[96] Anaximenes (?),[97] Archelaos (?),[98] Archigenes,[99] Aretaios,[100] Aristotle,[101] Caraka,[102]

---

94 Cf. also the index of work titles and the comparative chart in the appendix to the present book (pp. 203 and 208–214 below); indiscriminate allusions in the *Health Regimen* to 'Babylonian', Greek, Indian or Persian templates have not been included in the following list—see, for these, the index of people and places (p. 202 below).

95 A few more are thrown up, without references, in his Arabic introduction (pp. يه–يو).

96 E.g. ṬabFir 66,6 (called "the sage" [*al-ḥakīm*]); 144,12 (called "the Alexandrian [!] philosopher" [*al-failasūf… al-iskandarānī*]); elsewhere also known as "the traveller" (*aṭ-ṭauwāf*, calque of *περιπατητικός*). For certain, strictly medico-pharmaceutical passages, Alexander *of Tralles* may be meant.

97 ṬabFir 420,4 with note ٢.

98 ṬabFir 348,16 and *passim*.

99 ṬabFir 150,20 and *passim*.

100 ṬabFir 350,1.

101 ṬabFir 6,11 and *passim*: *Ἠθικὰ Εὐδήμεια*; 73,12 and *passim*: *Κατηγορίαι*; 21,24 and *passim*: *Μετεωρολογικά*; 13,14 and *passim*: *Περὶ γενέσεως καὶ φθορᾶς*; 32,14 and *passim*: *Περὶ ζῴων γενέσεως*; 76,5 and *passim*: *Περὶ ζῴων μορίων*; 20,17 and *passim*: *Περὶ οὐρανοῦ*; 114,21 and *passim*, § 24 with note 30, § 46 with note 54, § 99 with note 102: *Περὶ ψυχῆς*; 69,15 and *passim*: *Τὰ μετὰ τὰ φυσικά*; 35,2 and *passim*: *Τῶν περὶ τὰ ζῷα ἱστοριῶν*; 61,13 and *passim*: *Φυσικὴ ἀκρόασις*.

102 ṬabFir 557,11 and App2 *passim*, § 22 with note 27, notes 37, 70, 74–78, 85, 87–88, 91–94, 97, 99: *Saṃhitā*.

Democritos (Bolos),[103] Dioscorides,[104] Galen,[105] Hippocrates,[106] Ḥunain ibn Isḥāq,[107] Job of Edessa,[108] Mādhava,[109] Magnus of Emesa,[110] Māsarǧawaih,[111] Physiologos (?),[112] Plato,[113] Polemon of Laodicea,[114] ps-Aristotle,[115] ps-Galen,[116] Ptolemy,[117] Pythagoras,[118] Stephanos (Alexandrinos[?]),[119] Suśruta,[120] Theo-

---

103 ṬabFir 485,5.

104 App2 *passim*, [§216] note 113, §218 with note 117, §222 with note 118, §227 with note 119: *Περὶ ὕλης ἰατρικῆς*.

105 App2: *Εἰς τὸ Ἱπποκράτους ἐπιδημιῶν ὑπομνήματα*; ṬabFir 541,10: *Εἰς τὸ Ἱπποκράτους περὶ ἀέρων ὑδάτων τόπων ὑπομνήματα*; §47 with note 56: *Εἰς τὸ Ἱπποκράτους περὶ τροφῆς ὑπομνήματα*; App2 *passim*, §11 with note 16, §132 with note 104: *Θεραπευτικὴ μέθοδος*; App2: *Περὶ αἰτιῶν συμπτωμάτων*; §252 with note 126: *Περὶ ἀντιδότων*; App2, §242 with note 122: *Περὶ θηριακῆς πρὸς Πίσωνα*; App2, §12 with note 17, §47 with note 55: *Περὶ κράσεων*; App2 *passim*, §57 with note 58, §97 with note 100, §98 with note 101, [§216] note 113, §216 with note 116, §240 with note 121: *Περὶ κράσεως καὶ δυνάμεως τῶν ἁπλῶν φαρμάκων*; §47 with note 56: *Περὶ τῶν ἐν ταῖς νόσοις καιρῶν*; §36 with note 42: *Περὶ τῶν Ἱπποκράτους ἀφορισμῶν*; App2 *passim*: *Περὶ τῶν πεπονθότων τόπων*; App2: *Περὶ φλεβοτομίας πρὸς Ἐρασιστρατείους τοὺς ἐν Ῥώμῃ*; App2, §10 with note 14: *Ὑγιεινά*.

106 App2 *passim*, §37 with note 43, §215 with note 111: *Ἀφορισμοί*; App2, §15 with note 21, §32 with note 41: *Ἐπιδημίαι*; App2: *Κωακαὶ προγνώσιες*; App2 *passim*, §215 with note 111: *Περὶ ἀέρων ὑδάτων τόπων*; App2: *Περὶ ἄρθρων*; App2, §19 with note 26: *Περὶ ἀρχαίης ἰητρικῆς*; App2: *Περὶ γονῆς*; §18 with note 25: *Περὶ διαίτης (τὸ δεύτερον)*; App2, §16 with note 23, §17 with note 24, §19 with note 26: *Περὶ διαίτης ὀξέων*; App2: *Περὶ διαίτης ὑγιεινῆς*; App2 *passim*: *Περὶ ἑβδομάδων*; App2 *passim*: *Περὶ ἱερῆς νούσου*; App2: *Περὶ ἰητροῦ*; App2: *Περὶ κρισίων*; §29 with note 38: *Περὶ τροφῆς*; App2 *passim*: *Περὶ φύσιος ἀνθρώπου*; App2 *passim*: *Περὶ φύσιος παιδίου*; App2: *Περὶ φυσῶν*; App2 *passim*, §79 with note 69: *Προγνωστικόν*.

107 E.g. ṬabFir 8,10 (called "the translator" [*at-tarǧumān*]).

108 ṬabFir 347,20–354,17; §79 with note 69: *Kṯāḇā dTapšūrṯā*.

109 ṬabFir 557,12 and implicitly *passim*, §29 with note 37: *Nidāna*.

110 ṬabFir 351,17 and *passim*: *Περὶ οὔρων*.

111 ṬabFir 465,6 and *passim*.

112 ṬabFir 534,16 and *passim*: *Ṭabāʾiʿ al-ḥayawān*.

113 ṬabFir 16,1 and *passim*.

114 ṬabFir 49 note ٢ line 5, §74 with note 68: *Περὶ φυσιογνωμονίας*.

115 ṬabFir 4,1 and *passim*: *Περὶ κόσμου*. Relevant here is also the apocryphal *Sirr al-asrār* or "Secret of Secrets" which Ṭabarī, in his *Paradise of Wisdom* and by extension in the *Health Regimen*, too, cites without indicating a source, for example: BadUṣ 87ff., 91, 93ff., 96ff., 100ff. = §§10–15, 19, 33–38, 41–45, 48, 55–58, 94, 215 = ṬabFir, corresponding entries in the appendix to the present book (pp. 208–214 below).

116 ṬabFir 354,19–355,18; §79 with note 69: *Περὶ οὔρων*.

117 ṬabFir 547,11 and *passim*: *Γεωγραφικὴ ὑφήγησις*.

118 ṬabFir 65,12 and *passim*.

119 ṬabFir 144,12 and *passim*.

120 ṬabFir 557,11f. and App2 *passim*, §22 with note 27, notes 95, 97: *Saṃhitā*.

phrastos,[121] Vāgbhaṭa,[122] Vindanios Anatolios,[123] Yūḥannā ibn Māsawaih.[124]

Two further remarks in the *Paradise of Wisdom* may be worth noting: right at the beginning, Ṭabarī speaks of certain medical works (*kunnāšāt*), written by 'Syrians' (*ahl sūrīyā*), upon which he relied, among others, when composing his own chef-d'œuvre[125]—these works include, quite probably, the *Kunnāš al-Ḫūz*, a medico-pharmaceutical compendium assembled in Syriac by some physicians from Gondēšāpūr;[126] Job of Edessa's *Kṯāḇā ḏSīmāṯā* or "Book of Treasures", an encyclopedia of natural philosophy;[127] and the anonymous Syriac *Book of Medicines*.[128] Finally, towards the end of the *Paradise of Wisdom*, Ṭabarī gives various gynecological prescriptions which, he says, are taken "from the writings of an Indian woman" (*min kutub imra'a hindīya*),[129] but which are impossible to link to any documented Ayurvedic source.

---

121 ṬabFir 63,7 and *passim*.

122 ṬabFir 557,12 and implicitly *passim*, § 22 with note 27, notes 70, 74–77, 79, 81, 85, 87–88, 91–93, 95, 97, 99: *Aṣṭāṅgahṛdayasaṃhitā*.

123 ṬabFir 518,1 and *passim*: *Συναγωγὴ γεωργικῶν ἐπιτηδευμάτων*.

124 E.g. ṬabFir 8,9 (called Yūḥannā *bar* [!] Māsawaih "the physician of the king" [*ṭabīb al-malik*]); 463,5 (called Yaḥyā).

125 See, for the full passage, ṬabFir 1,17–2,3.

126 This compendium seems to be cited also explicitly in ṬabFir 476,25–477,1 s.t. *Kitāb a(l-)Hūz* (اهوز sic); cf. for this work RhaCB 36–42 (s.n. Hūzāyē).

127 This encyclopedia was identified as a source of the *Paradise of Wisdom* by Alphonse Mingana, see JobBT xxvi.

128 A connection between the Syriac *Book of Medicines* and the *Paradise of Wisdom* was established, secondarily, by Max Meyerhof, see MeyṬab 62 f.

129 See ṬabFir 591,10 for the explicit citation and 591,11–594,13 for the prescriptions. Perhaps this woman is identical with an-Nadīm's Rwsā (lege *Rūšī* < Sanskrit *Ruci* [?]), an Indian female (*hindīya*) who is said to have authored a "Book on the Treatments of Women" (*Kitāb ... fī ʿIlāǧāt an-nisā'*), see NadFih 1/303,11 with 2/147,14–17 (Arabic variants for the name: دوينى / روبسى / روشى / روسا). It may be noted here in passing that there are no correspondences between the Indian material in the *Paradise of Wisdom* and the passages which Masīḥ ad-Dimašqī (d. after 225/840), otherwise one of Ṭabarī's minor and *implicit* sources (cf. p. 98 below note 68), links in his *Risāla Hārūnīya* to the teachings of an Indian (astrologer-physician?) called *Flṭys* (فلطيس), whom he often cites in one breath with well-known *Greek* authorities; for these passages see DimRH 41, 65, 77, 81, 87, 91, 97, 105, 115, 119, 175, 179, 181, 199, 201 and 439. Assuming that *Flṭys* is not an altogether fictitious personage, it seems likely that he served Masīḥ as an *oral* informant during his stay in India, cf. also Gigandet's assessment in DimRH 17 and 40 note 2; on a sidenote I should like to remark that there is no justification to transliterate the Arabic form of this unidentified Sanskrit name as *Falaṭīs*. Pretty much the same goes for *'mqt* (امقت), another Indian whom Masīḥ encountered at the court of Hārūn ar-Rašīd (reg. 170/786–193/809), and whom he allegedly

c Discussion

In the preceding sections, I have tried to place the text edited and translated here into a reasonably detailed historical frame—from its genre-specific background, through a biobibliographical account of its author, down to its manuscript manifestations and distinctly identifiable sources. In the present and final introductory section, we will deal more broadly with the text's compositional elements, its relationship to the *Paradise of Wisdom*, its particular handling of authoritative information and notably its sweeping references to Greek, Indian and 'Babylonian' templates, its general literary character and self-declared destination, its wider cultural setting, as well as its documentary significance for us.

First, however, we need to return briefly to the two Arabic manuscripts which contain the text of the *Health Regimen*, as they have been subject to certain misconceptions on the part of some modern researchers. The Oxford manuscript offers the complete text under the title *Fī Ḥifẓ aṣ-ṣiḥḥa* "On the Preservation of Health"; the Istanbul manuscript offers the first half of the text under the title *Kitāb al-Luʾluʾa* "Book of the Pearl".[130] Manfred Ullmann and Fuat Sezgin, none of whom appears to have had any close familiarity with either manuscript, invoke an untraceable statement by Max Meyerhof, to the effect that the two manuscripts, once properly examined, might turn out to be identical.[131] The German translator of the *Health Regimen*, Usama Raslan, writing a few years later, established this assumption as a fact[132]—a mere glance at the apparatus to the present edition will easily corroborate his conclusion. Elvira Wakelnig's assertion that "the *Kitāb al-Luʾluʾa* is a shortened version of the first half of the *Kitāb Ḥifẓ al-Ṣiḥḥa*, either condensed by ʿAlī ibn Rabban himself or, probably more plausible, by some later redactor" is therefore inaccurate.[133] The text preserved in the Istanbul manuscript, albeit slightly pruned,

accompanied on his return journey to India, see DimRH 333; a panacea named "the helper" (*al-muġīṯ*), whose composition *ʾmqt* left behind, is also recorded by Masīḥ (DimRH 335) but untraceable in Ayurvedic literature.

130 For a concise description of these manuscripts see p. 14 above, where I also already indicated that the titular discrepancy may well be explained by later, non-authorial splittings of a supposed original title along the lines of **Kitāb al-Luʾluʾa fī ḥifẓ aṣ-ṣiḥḥa** "Book of the Pearl—on the Preservation of Health" (only one medieval Arab biobibliographer has at all retained a memory of the work, namely IAU 2.2/759 no. 6 s.t. *Kitāb Ḥifẓ aṣ-ṣiḥḥa*).

131 UllMed 190 and, respectively, GaS 3/239 note 1 (both published in 1970).

132 ṬabHT 8,3 (published in 1975).

133 WaṬab 219; a similar judgement is passed by Olsson, cf. OlDDS 27. Both Wakelnig and Olsson seem to have been misled by certain (ambiguous) remarks made by the German translator—thus, Raslan contradicts himself when he says: "Das *K. al-Luʾluʾa* ist vollständig [!] in der ersten Hälfte des *K. Ḥifẓ aṣ-ṣiḥḥa* enthalten" (ṬabHT 14,3 f.), only to declare a

is neither an abbreviation nor a condensation of the text preserved in the Oxford manuscript, but simply a version from which the second half has been omitted altogether; the questions as to what motivated this amputation of the text—and whether that decision was taken by the copyist, by a reviser or even by a hypothetical commissioner—are impossible to answer without digressing into the realm of speculation. It is important to note that these considerations have no bearing on the literary characterization of the *Health Regimen* as a whole, which will be discussed further on.

In order to arrive at a fair assessment of the relationship between the *Health Regimen* and the *Paradise of Wisdom*, it may be useful to start with a summary of the chapters into which the former text is divided. In the manuscript tradition, the *Health Regimen* carries the following chapter-headings:[134]

[Introduction]	–	§§1–9
On the Preservation of Health	*Fī Ḥifẓ aṣ-ṣiḥḥa*[135]	§§10–19
On nourishments and (other) measures which strengthen, fatten, thin, arouse or appease the body	*Fī l-aġḏiya wat-tadābīr allatī tuqauwī l-badan wa-tusammin wa-tuhazzil wa-tuhaiyiǧ wa-tusakkin*	§§20–26
On the damages resulting from insomnia, passion, excessive purging, sex addiction, and overmuch bloodletting—it is because of this chapter that we wrote the book	*Fī l-maḍarrāt al-ḥādiṯa min as-sahar waš-šahwa wa-farṭ al-ishāl wa-idmān al-bāh wa-kaṯrat iḫrāǧ ad-dam wa-huwa l-bāb allaḏī allafnā l-kitāb li-aǧlihī*	§§27–32
On how to look after the body in spring	*Fī tadbīr al-badan fī r-rabīʿ*	§33
On how to look after the body in summer	*Fī tadbīr al-badan fī ṣ-ṣaif*	§34

few lines later: "Da das *K. al-Luʾluʾa* bei den Biographen nirgends faßbar ist, muß man annehmen, daß es sich hier um eine abgekürzte [!] Fassung des *K. Ḥifẓ aṣ-ṣiḥḥa* handelt" (ibid. 14,-4ff.). It may be remarked here in passing that the Istanbul manuscript, despite being stylistically more sloppy, has preserved an illustration which is missing from the Oxford manuscript, see §67 of my edition/translation of the text.

134 Considering the idiosyncratic (if not to say arbitrary) positioning and phrasing of these captions, it seems likely that at least some of them, in their given forms, are non-authorial insertions made by subsequent copyists of the work.

135 This caption also indicates one of the two titles of the work.

(*cont.*)

On how to look after the body in autumn	*Fī tadbīr al-badan fī l-ḫarīf*	§35
On how to look after the body in winter	*Fī tadbīr al-badan fī š-šitā'*	§§36–38
On how to look after the body in the bathhouse	*Fī tadbīr al-badan fī l-ḥammām*	§§39–40
On symptoms and treatments (according to) Babylonian and other physicians	*Fī ʿalāmāt wa-ʿilāǧāt aṭibbā' bābil wa-ġairihim*	§§41–45
On the reason for self-nutrition	*Fī ʿillat al-iġtiḏā'*	§§46–47
On the values and benefits of food (items)	*Fī aqdār al-aġḏiya wa-manāfiʿihā*	§§48–53
On milks	*Fī l-albān*	§54
On wines	*Fī l-ašriba*	§55
On oils and others	*Fī l-adhān wa-ġairihā*	§§56–58
On tender and on tough nourishments	*Fī l-aġḏiya al-laṭīfa wal-ġalīẓa*	§§59–60
On the powers of (humoral) mixtures, the reasons for their agitation, the causes of their generation, and clear evidence for all that	*Fī quwā l-mizāǧāt wa-ʿilal hayaǧānihā wa-asbāb taulīdihā wad-dalā'il al-wāḍiḥa ʿalā ḏālika kulluhū*	§§61–68
On the reasons for the agitation and predominance of temperaments and mixtures	*Fī ʿilal hayaǧān aṭ-ṭabā'iʿ wal-mizāǧāt wa-ġalabatihā*	§§69–72
On how to discern (specific) mixtures and their preponderance over the body	*Fī d-dalīl ʿalā mizāǧāt al-abdān wal-ġālib ʿalaihā*	§§73–79
The teachings of the Indians about health regimen	*Qaul al-hind fī tadbīr aṣ-ṣiḥḥa*	§§80–96
On knowing the powers of substances and on inferring them from their taste—after the teachings of the Greeks	*Fī maʿrifat quwā l-ašyā' wal-istidlāl ʿalaihā bi-maḏāqatihā min qaul ar-rūm*	§§97–116
On greenstuffs and vinegar-based pickles	*Fī l-buqūl wal-kawāmīḫ*	§§117–156
On fruits and fruity edibles	*Fī l-fākiha waṯ-ṯimār*	§§157–184
Aromatic plants	*ar-Rayāḥīn*	§§185–206

(*cont.*)

Preserves	*al-Anbağāt*	§§ 207–214
Waters	*al-Miyāh*	§ 215
On the useful properties of animal parts	*Fī manāfiʿ aʿḍāʾ al-ḥayawān*	§§ 216–246
[Compound Drugs]	–	§§ 248–261

Broken down into perhaps more palatable compositional units, the topical range of the text can be paraphrased as follows:

Prolegomena	§§ 1–9
General Dietetics and Prophylactic Hygiene (Greek)	§§ 10–19
Dietary Regimen, Sexual and Somnial Hygiene, Psychosomatics, Phlebotomy and Catharsis	§§ 20–32
Seasonal Regimen and Personal Hygiene	§§ 33–40
Basic Physiopathology ('Babylonian')[136]	§§ 41–44
Daily Regimen	§ 45
Nutrition and Decline	§§ 46–47
Aliments and Alimentation	§§ 48–60
Principles of Humoral Physiology	§§ 61–67
Superstructure of Humoralism	§ 68
Principles of Humoral Pathology	§§ 69–72
Physiognomy and Uroscopy	§§ 73–79
General Hygiene, Daily Regimen and Personal Conduct (Indian)	§§ 80–96
Gustology (Greek)	§§ 97–100
Materia Medica	§§ 101–215
Organotherapy and Sympathetic Medicine	§§ 216–246
Medical Formulas	§§ 248–261

A comparison between the formal divisions of Ṭabarī's *Health Regimen*, as outlined above, and those he employed for the creation of his *Paradise of Wisdom*,[137] shows that the structure of the two works is radically different. This is no surprise when considering that the *Health Regimen* is a relatively short,

136 On the concept of 'Babylonian' see p. 26 below.

137 It would be tedious to reiterate here the numerous divisions and subdivisions of the *Para-*

straightforward medical manual, as opposed to the *Paradise of Wisdom*, which is an extensive medico-philosophical encyclopedia with a distinctively hybrid complexion. Whilst Ṭabarī's authorship of the *Health Regimen* is contested neither by the old Arab copyists nor by the few modern scholars who voiced an opinion on the subject, the work itself has been defined as an abridgement or extract of the *Paradise of Wisdom*, most notably by Usama Raslan, the German translator of the text, who obviously was very well acquainted with it;[138] he goes on to profess his 'disappointment' that the *Health Regimen* is all too closely related to the *Paradise of Wisdom* and therefore offers hardly any additional material.[139] Statements like this, however, misappreciate the nature of the problem: Raslan's judgement is valid only on a quantitative level, insofar as it aims at describing the raw informational content of the work; in reality, the *Health Regimen* is much better characterized as a redesigned abstract of the *Paradise of Wisdom*, written for a different, narrower target group and focused exclusively on topics relevant to hygienics or preventive medicine. It is stripped of any theoretical accessories and medico-philosophical technicalities, except for some very basic excursions into the micro- and macrocosmic surroundings of humoralism;[140] on the other hand, there are numerous explanatory digressions and instructive amplifications,[141] as well as some embellishments in the *adab* vein, like anecdotes[142] and even a little poem;[143] and there is a heedful emphasis on the 'universality' of conveyed information, to which "the scholars of all races" (*al-ʿulamāʾ min al-umam kulluhā*) have subscribed.[144] Whilst the overall style of writing testifies to the same authorship, the general layout

---

dise of Wisdom, so suffice it to refer the reader to the readily available, highly detailed chapter listings given by Ṣiddīqī (in Arabic) and Meyerhof (in English), see ṬabFir (20 unpaginated pages following immediately on the editor's preface to his edition) and, respectively, MeyPW 17–46; cf. also BrowAM 42 f. for a synopsis.

138 Raslan qualifies the *Health Regimen* as "ein Auszug aus dem *Paradies der Weisheit*" (ṬabHT 8,7 f. and 15,5); Ebied and Thomas provisionally call it "an abridgement" (EbThPW 14 note 49). Emilie Savage-Smith, who catalogued the Oxford manuscript with great diligence, prudently speaks of "similarities between this treatise and portions of the *Kitāb Firdaws al-ḥikmah*" (SSNC 563,-3 f.).

139 ṬabHT 13,7–12. For curiosity's sake, it may be added here in passing that Otto Spies, in his short review of Raslan's translation, propounded the strange opinion that the *Health Regimen* is a conglomerate of extracts ("Auszüge" [!]) which, not Ṭabarī but "ein Schüler oder späterer Mediziner […] für sich veranstaltet hat" (SpiWG 256).

140 Cf. §§ 62–68.

141 For example §§ 74–78.

142 See §§ 4, 5, 6, 25, 26, 45 (*supra*), 53 (*infra*).

143 See § 3 (*infra*).

144 See § 3 (*supra*) and *passim*; cf. also § 8 (*supra*).

of both works is not at all the same. The comparative chart in the appendix to the present book[145] illustrates that in terms of factual content, about 80 % of the *Health Regimen* are drawn directly from the *Paradise of Wisdom*; but it also shows, first, that the topical sequence in the two texts has been altered completely by Ṭabarī and, second, that literal borrowings are the exception, whereas reformulations, simplifications and paraphrases are the rule.[146] This means that even in those cases where the same item of information is communicated, Ṭabarī adjusted it terminologically and remoulded it stylistically, so as to suit the needs and expectations of an audience of educated nonprofessionals; and it also means that the *Health Regimen*, through whichever lense of literary criticism it may be viewed, is a creation conceivably at variance with the *Paradise of Wisdom*. This reflection notwithstanding, the latter work can justly be regarded as something like a mother-text when it comes to the next question, namely that of Ṭabarī's sources and, more specifically, his handling of them.

In the *Paradise of Wisdom*, Ṭabarī names his sources neither frequently nor consistently, and this deliberate restraint is carried over into the *Health Regimen*, where only four authorities, all Greeks, are mentioned by name: Aristotle,[147] Dioscorides,[148] Galen[149] and Hippocrates.[150] There are, to be sure, several other sources that underlie the *Health Regimen* and that can be identified indirectly;[151] there are instances in which Ṭabarī tacitly transmits information from the very sources he otherwise openly names;[152] and there are collective references to Greek, Indian, 'Babylonian' and, in one case, Persian sources, which we will deal with shortly. Identifying individual sources is one thing, analyzing the way they are handled is another. Few scholars have so far made some preliminary remarks on this thorny issue, departing from Ṣiddīqī's Arabic edition of the *Paradise of Wisdom*, and there focusing on Ṭabarī's implementation

145 See pp. 208–214 below.

146 As the chart only captures positive data, non-entries are to be considered 'new' material (c. 20 %).

147 See §§ 24, 46, 99.

148 See §§ 218, 222, 227.

149 See §§ 10, (11), (12), 14, 36, 47, 57, 97, 98, 216, 240.

150 See §§ 15, (16), 17, (18), 19, 29, 32, 37, 215.

151 For a detailed, albeit provisional, list of sources in the *Paradise of Wisdom* and, by proof or implication, in the *Health Regimen*, see pp. 15 ff. above.

152 This is true in particular for §§ 102–214, which are concerned with the properties of simple substances—much of the material exhibited in this section of the *Health Regimen* has been imported from Dioscorides' *Περὶ ὕλης ἰατρικῆς* (cf. UllMed 121,-6 ff.) and, to a lesser extent, from Galen's *Περὶ κράσεως καὶ δυνάμεως τῶν ἁπλῶν φαρμάκων*.

of *Greek* material. Thus, Max Meyerhof, writing in 1931, observed that Ṭabarī's 'quotations' from Greek authors are generally imprecise ("ungenau") and probably based on Syriac intermediate translations, rather than on already existing Arabic translations made by Ḥunain ibn Isḥāq (d. 260/873) and his circle;[153] he moreover pointed out that substantial segments of the *Paradise of Wisdom* can be traced to the anonymous Syriac *Book of Medicines*,[154] which was never translated into Arabic. Manfred Ullmann, writing in 1970, and some others in his wake basically say the same.[155] A systematic and fairly recent investigation into Ṭabarī's use of specific Greek sources was conducted in 2015 by Joshua Olsson, but even though his discussion of the problem is more sophisticated and more nuanced than that of his predecessors, he in the end arrives, albeit with some caveats, at very similar conclusions.[156] As regards *Sanskrit* material in the *Paradise of Wisdom*, no serious studies have been made to date. Ṭabarī's Indian sources (Suśruta's *Saṃhitā*, Caraka's *Saṃhitā*, Vāgbhaṭa's *Aṣṭāṅgahṛdayasaṃhitā* and Mādhava's *Nidāna*)[157] had been translated, in Baghdad, from Sanskrit (sometimes through Pahlavi) into Arabic between, roughly, the years 786 and 820 CE.[158] Alfred Siggel, who in 1950 translated large parts of the so-called 'Indian Books'[159] into German, emphasized that Ṭabarī reproduced these texts mostly in the form of free renditions, seldom literally, often altered or abridged.[160] And Werner Schmucker, writing in 1969, who knew Sanskrit bet-

153 "Die Zitate aus dem Griechischen sind fast immer ungenau und machen den Eindruck, als ob sie aus dem Syrischen übersetzt worden wären. Jedenfalls entstammen sie nicht den von Ḥunain und seinen Schülern geschaffenen arabischen Übersetzungen der Werke des Hippokrates und Galenos, sondern syrischen Übersetzungen, aus denen er [scil. Ṭabarī] selbst in das Arabische frei übersetzt hat" (MeyṬab 62).

154 MeyṬab 63; similarly, Alphonse Mingana, writing in 1935, established that several passages in the *Paradise of Wisdom* have been borrowed from Job of Edessa's (d. c. 220/835) Syriac *Book of Treasures* (JobBT xxvi with note 2).

155 "Bemerkenswerterweise zitiert 'Alī [aṭ-Ṭabarī] die griechischen Autoren noch nicht nach den Übersetzungen der Schule Ḥunains, sondern in freien arabischen Wiedergaben syrischer Vorlagen" (UllMed 122); regarding Dioscorides' *Materia Medica*, Ullmann further notes that "der Autor [scil. Ṭabarī] führt die Stellen nicht nach der arabischen Übersetzung des Iṣṭafān [fl. mid 3rd/9th century] an, sondern übersetzt vermutlich selbst aus der syrischen Vorlage" (ibid. 258).

156 OlDDS 32–66.

157 Explicitly named ṬabFir 557,11 f. s.nn. Susrud, Ǧarak, *Aštānqahrday* and *Nidān*.

158 See RhaCB 17, 19, 22 and 27; cf. also the slightly updated chart in KaIMT 96 f.

159 ṬabFir 557–620.

160 "Der Vergleich der Darlegungen R.T.s [scil. Rabban aṭ-Ṭabarī's] mit den uns zugänglichen Übersetzungen jener indischen Ärzte zeigt, daß die 'Indischen Bücher' im *Paradies der Weisheit* einen freien Auszug darstellen, in dem nur einige Stellen wörtlich wiedergegeben werden, andere in veränderter und gekürzter Form erscheinen" (SiIB 1102 = [8]).

ter than Siggel did, believed, in my opinion on good grounds, that Ṭabarī was accessing Ayurvedic material through Pahlavi (rather than Syriac) intermediate translations.[161] Ṭabarī's handling of transmitted information with, by our standards, considerable liberality and authorial freedom was by no means uncommon in his day and age—it reflects, on the contrary, a rather typical intellectual attitude found also in other medico-philosophical compilations (and beyond). We still lack sufficient data, but it is clear that, for example, the physician Abū Bakr ar-Rāzī (d. 313/925), in his *Comprehensive Book*, proceeded in exactly the same way with his own sources, subjecting them freely to amplifications, condensations, abridgements, rearrangements, insertions, omissions, fusions, reformulations, stealth comments, and so on.[162] There are, in theory, four strands of literary sources that would have been used by Ṭabarī for the composition of the *Paradise of Wisdom*: Greek, Sanskrit, Syriac and Pahlavi. It is likely that he 'quoted' Greek texts mainly through (now lost) Syriac and occasionally through Arabic translations, whilst Sanskrit texts were available to him in either (now lost) Pahlavi or Arabic renditions; and it seems certain that his management of these sources in the *Health Regimen* was no different from what he had done in the *Paradise of Wisdom*. All source 'verifications', as they will be found in the footnotes to my translation of the *Health Regimen*, are therefore set against (and need to be interpreted on) the background of the foregoing considerations—namely, as rough guides to distant prototypes, and as examples of textual change rather than illustrations of transmissional stability. To conclude the subject of sources, it remains to say a few words about 'group references' in the *Health Regimen*. These references attribute clusters of information *collectively* to physicians whom Ṭabarī identified as Greeks,[163] Indians[164] or 'Babylonians';[165] far less prominently, and restricted to smaller items of information, he invokes or implies Persians and Byzantines.[166] As regards

161 "Daß ihm [scil. Ṭabarī] syrische Übersetzungen indischer Mediziner vorgelegen haben […] scheint mir fragwürdig. Persische Übersetzungen kommen durchaus in Frage" (SchṬab 41).

162 For Ṭabarī's 'editorial techniques' see OlDDS 47–61; for Rhazes' approach see BryKḤ 21 and 23–73 with WeiZit 281ff. (Greek sources), then RhaCB 7–12 (Sanskrit sources).

163 See §§1, 2, 80 and 101 (with note 103); further the chapter-heading prior to §97.

164 See §§1, 2, 22, 29, 45, 80 and 260; further the chapter-headings prior to §§41 (with note 48) and 80.

165 See §1 and the chapter-heading prior to 41.

166 On one occasion, Ṭabarī relates an anecdote on the authority of what appear to be Persians, featuring also a physician from Byzantium (§45 [*supra*] with note 49); the Persians and the Byzantines further seem to be involved in the transmission of certain physiopathological explications (§§41–44 with note 48); and in the second half of §45, the Persians are explicitly linked to some semi-occult dietary instructions, whose origin is likely to be oral. Note that the Persians, when named expressly, are called *al-furs*, the Byzantines *ar-*

the Greeks (*ar-rūm*), they obviously include any number of relevant authorities mentioned earlier on, and here notably Galen, Hippocrates and Dioscorides;[167] as for the Indians (*al-hind*), they are represented, just like in the *Paradise of Wisdom*, by Suśruta, Caraka, Vāgbhaṭa and Mādhava.[168] Less straightforward is the question of who are the 'Babylonian' scholars and physicians (*'ulamā'/ aṭibbā' bābil*). Medieval Arab authors used the term Bābil mainly to denote the region or 'climate' that more or less corresponds to Iraq, but in some sources[169] the notion of what constitutes Bābil is narrower, and restricted to the arable stretches of *southern* Iraq (*as-sawād*); in the latter sense, Bābil is invariably associated with the land of the Chaldeans on the lower Euphrates and Tigris.[170] In early Islamic times, Bābil became the name of an administrative district and, until about 900 CE, also the name of this district's capital, which was situated circa 60 miles due south of Baghdad, off the road to Kufa;[171] by the end of the 10th century CE, the city had turned into a small village.[172] In Ṭabarī's days, Bābil may well have been a thriving and perhaps even exciting town, but it seems much more likely that for him, too, Bābil was the whole of southern Iraq, including Kufa, Wasit, and probably Basra. Which (group of?) 'scholars and physicians' from that quarter Ṭabarī actually had in mind, remains impossible to say.

Any attempt at describing the literary character of the *Health Regimen* must take into account the purpose which that text explicitly professes to serve. Right at the beginning, Ṭabarī declares that he undertook its composition out of a sense of devotion and gratitude towards his benefactor (*mun'im*), the Commander of the Faithful, whom he also refers to as his patron (*saiyid*) and protector (*maulan*); he makes no secret that a wish to retain his own status (*'izz*) might have played a role as well.[173] From our preceding biographical survey it seems certain that the 'Commander of the Faithful' to whom Ṭabarī addresses the work is no other than the 'Abbāsid caliph al-Mutawakkil (reg. 232/847–247/861)—the man who witnessed (and possibly instigated) Ṭabarī's

---

rūm; the latter term is normally used by Ṭabarī to denote the (ancient) Greeks, and I have translated it depending on the context. Sweeping references to "sages" (*ḥukamā'* [§§67, 74, 75]) are hard to substantiate.

167 See p. 16 above.

168 Cf. pp. 15 ff. above.

169 For example YāBul 1/447,22 f.

170 Cf. CRCha 662a.

171 See LeSEC map 2.

172 On Bābil in Arabic literature still see the excellent article (from 1913!) by Ernst Herzfeld, containing virtually all relevant source indications (HeBā, esp. 571a).

173 See §1.

conversion to Islam soon after the year 235/850, and who subsequently received him into his inner circle.[174] Having paid lip-service to the notion of prophetic and, by extension, caliphal omniscience, Ṭabarī goes on to say that the *Health Regimen* is designed to be a reminder and a guide (*taḏkira wa-miṯāl*), first and foremost, for the caliph and for his princely sons, but also for anyone else who values knowledge[175]—we may take the latter aspiration to be a conventional figure of speech. After a preliminary and very basic discourse on the merits of moderation and the importance of proper nutrition, loosened up by some entertaining episodes, Ṭabarī brings his prolegomena to a close by reassuring the reader that the book is written in plain language, without pretence (*talbīs*) or profundity (*ġawīṣ*), and in an adequate, easy and uncomplicated style (*bi-waṣf muqniʿ yasīr ġair ʿasīr*).[176] In other words, the *Health Regimen* was conceived as a personal *vademecum* for some lofty members of the royal household and not, like the *Paradise of Wisdom*, as a high-aiming, far-reaching, all-embracing medico-philosophical encyclopedia. Ṭabarī's immediate and declared target group consisted of the caliph himself and his offspring, and in this respect, too, the *Health Regimen* is radically different from the *Paradise of Wisdom*, which latter represents the sum of a lifelong striving for knowledge and insight, congealed between two covers and addressed to the world. The 'private' nature and presumably limited circulation of the *Health Regimen*, combined with the fact that it contains relatively few items of information that are not already found in the *Paradise of Wisdom*, explain why Ṭabarī's small work has left no traces in subsequent Arabic medical literature. Now, it is certainly true that neither the caliph nor his sons were uneducated people—they were, on the contrary, well-tutored and well-versed in those branches of learning that would combine to bestow 'culture' on a man, and provide him with the skills necessary to perform given social and political functions, especially in high office; these branches of learning, in the days of al-Mutawakkil, included the Koran and prophetic traditions, Arabic grammar, poetry and prose, rhetoric, history, genealogy and ancient tribal lores, possibly some law and perhaps even, at least in theory, a smattering of Greek philosophy, Indian fables and Persian wisdom literature.[177] But however educated al-Mutawakkil may have been, he was no savant and no scientist, nor indeed acquainted with the principles of medicine—which is why Ṭabarī, when composing the *Health Regimen*, had to avoid technical jargon, construct a didactic narrative, cushion it at intervals

174 Cf. p. 8 above.
175 See § 2.
176 See § 9.
177 Cf. GabAd *passim*.

with entertaining or astounding side notes, and make it as palatable as possible for a perhaps interested but not necessarily enthusiastic lay audience;[178] besides, he had to make sure that his advice is practical, practicable and self-explanatory. It almost goes without saying—and is amply confirmed by even a casual glance over the text—that the *Health Regimen* was written for an affluent, leisured class of individuals, who could afford the recommended levels of medical care, and who had the means to choose from, and to procure, such a variety of dietary and therapeutic agents.[179] In a wider sense, and insofar as Arabic literature is concerned, Ṭabarī's *Health Regimen* can therefore be considered, not only the earliest extant text on hygienics but also the earliest extant representative of the memento or guidebook (*taḏkira*) genre.[180]

Ṭabarī's literary creations emerged from a rich soil of cultural diversity and intellectual vigour, and they need to be seen in the context of the great currents of the time. In the middle of the 3rd/9th century, the translation movement from Greek into Arabic had reached its apogee, and almost all Greek philosophical and scientific writings were gradually but swiftly becoming available to Arabic- or Syriac-speaking scholars;[181] the translations of several important Indian astronomical and, notably, medical works from Sanskrit into Arabic had already been accomplished about two generations earlier.[182] The reign of al-Mutawakkil, who tried to reassert the supremacy of the caliphate, was, despite ongoing social and political upheavals, a period of relative peace and stability, at least in the centres of power; the anarchy and the decline into which the caliphate was plunged after al-Mutawakkil's death in 247/861 caught Ṭabarī

178 From this point of view, the long section paraphrasing "the teachings of the Indians about health regimen" (§§80–96) may be interpreted to serve the double purpose of conveying hygienic information, and of providing intellectual stimulation: quite a few prescriptions and rules of conduct recounted in this section are culturally so specific that they would have made little sense to a Muslim sitting in Baghdad or Samarra, other than as objects of curiosity and wonder.

179 For these see §§102–214 and *passim*.

180 We have already seen that Ṭabarī himself, in §2, calls his book a "reminder and guide" (*taḏkira wa-miṯāl*). A lost precursor of the medical *taḏkira* genre is the *Kitāb at-Taḏkira*, which Buḫtīšūʿ ibn Ǧūrǧis (d. 185/801) wrote for his son, see IAU 2.1/344 no. 2 with GaS 3/211. The list of genre successors that deal with general medicine is long, for example Abū Saʿīd ʿUbaidallāh Ibn Buḫtīšūʿ's (d. after 450/1058) *Taḏkirat al-ḥāḍir wa-zād al-musāfir*, of which *ar-Rauḍa aṭ-ṭibbīya* is an extract, see UllMed 110 with note 7; Abū l-ʿAlāʾ Zuhr's (d. 525/1131) *Kitāb at-Taḏkira*, written for his son, see UllMed 162 with note 3; down to Dāwūd al-Anṭākī's (d. 1008/1599) *Taḏkirat ulī l-albāb wal-ǧāmiʿ lil-ʿaǧab al-ʿuǧāb*, see UllMed 181 with note 5.

181 On the transmission of Greek texts to the Arabs see GuGT 1–8 and *passim*.

182 On the transmission of Sanskrit texts to the Arabs see PinHay 1136a–b (astronomy) and RhaCB 7–28 with KaIMT *passim* (medicine).

only at the tail-end of his life, and the eventual collapse of caliphal authority even at the core of the empire was yet to come.[183] The Banū Mūsā—wealthy geometricians, mechanics and sponsors of scientific translations—were key figures in the intellectual life of Baghdad; the philosopher and polymath al-Kindī, the mathematician Ṯābit ibn Qurra, the astrologer Abū Maʿšar, the astronomer al-Farġānī, the translator-physician Ḥunain ibn Isḥāq, the pharmacist Sābūr ibn Sahl, the historian al-Yaʿqūbī, the philologist al-Mubarrad, the grammarian Ibn as-Sikkīt, the famous prose writer al-Ǧāḥiẓ, the celebrated poets Abū Tammām and al-Buḥturī, to name but a few, were all active in Iraq and contemporaries of Ṭabarī. In a milieu such as this, a small work like the *Health Regimen* is bound to fade into the background, both in terms of scope and ambition. There is no doubt in my mind that Ṭabarī was a match for those illustrious fellows, and that his *Paradise of Wisdom*, with its profound depth of insight and sweeping breadth of knowledge, is a prime example of how heavily he could weigh in on the medico-philosophical debates of the day. The wider scientific community had little choice but to acknowledge and, in one way or another, to engage with the *Paradise of Wisdom*; on the other hand, the *Health Regimen*, due to its very design and direction, was probably not even known to contemporary scholars or, if it was, they would likely have considered it an insipid exercise on the part of Ṭabarī, a shrunken and simplified variation of the underlying masterpiece.

What documentary value, then, has a historical text like the *Health Regimen* for us? It has already been said that dismissing the *Health Regimen* as a mere abridgement or extract of the *Paradise of Wisdom* is misjudging both its general literary character and its specific social function.[184] Ṭabarī's medieval colleagues, coeval and posterior, seem to have largely ignored his *Health Regimen*—it is quoted nowhere in subsequent Arabic literature, and if we had not two lucky manuscripts of it at our disposal, we would know nothing about it except its title, which *one* diligent bibliographer has laconically recorded;[185] the apparent indifference of Ṭabarī's peers towards the *Health Regimen* may be explained either by a complete ignorance of its existence, or as the consequence of a pragmatic probing of its informational content vis-à-vis the *Para-*

183 Cf. KeMut *passim* and LeʿA 18b.

184 Cf. p. 22 f. above.

185 IAU 2.2/759 no. 6 (7th/13th century). The fact that two manuscripts of the *Health Regimen* have come down to us at all, shows that the text did attract the attention (or curiosity) of at least some medieval Arab scribes; however, these isolated instances date from about five centuries after the text's composition, and can hardly be regarded as tokens of ponderable impact.

dise of Wisdom, which would have resulted in disappointment. It is our task, I would argue, to view the *Health Regimen* in another perspective, and from an angle that disregards the sober question of medico-philosophical data increase. Seen and interpreted above all as a literary (rather than scientific) document, Ṭabarī's small work considerably gains in significance: it is the oldest surviving Arabic specimen of both the hygienics and the memento genre; it is the earliest extant Arabic medical text written for a lay audience; its largely non-technical lexicon and its expressive style combine to create an intersection with classical Arabic prose literature, and testify at the same time to the linguistic possibilities available during a formative period of intellectual conception; and finally, it must be recalled that the author of the *Health Regimen* was, by any standard, one of the most remarkable and most original thinkers of the 3rd/9th century. With all this in mind, we should be in a good position to do justice to Ṭabarī's small but highly unusual writing.

d *Metrological Units*

The weights and measures occurring in Ṭabarī's *Health Regimen* can be divided into three categories:[186]

	Specific	
dāniq	0.63 g	[§§ 252, 259]
dirham	3.13 g	[§§ 248 ff., 253 f., 258]
ḥabba	0.05 g	[§ 252]
miṯqāl	4.46 g	[§§ 45, 233, 254, 260]
raṭl	406 g	[§§ 256, 258, 260]
ūqīya	33 g	[§§ 256, 260]

Semispecific	
bowlful (*sukurruǧa*)	[§ 259]
broad bean (*bāqillāh*)	[§§ 98, 228, 230, 257]
chickpea (*ḥimmaṣa*)	[§§ 257, 260]
cupful (*qadaḥ*)	[§§ 95, 258]
finger, little (*ḫinṣir*)	[§ 81]
lentil (*ʿadasa*)	[§ 257]
morsel (*luqma*)	[§ 45]
sliver (*nuḥāta*)	[§ 218]

186 For 'Islamic' units, basic conversions and comparative data (largely drawn from non-medical sources) see HiMG *passim*; further UllMed 316–320 and SābAq 225–228 with the literature quoted in either section (adding KaWM).

span (*fitr* or *šibr*)	[§§ 25, 81]
spoonful (*milʿaqa*)	[§ 232]

Unspecific

amount (*zina*)	[§§ 249 f., 253]
dose (*šurba* or *wazn* or *zina*)	[§§ 248 ff., 252, 254, 257 ff.]
part (*ǧuzʾ*)	[§§ 248, 252, 255, 257, 259, 261]
quantity (*qadr*)	[§§ 228, 230, 232 f.]
size (*qadr*)	[§§ 98, 257]
weight (*wazn* or *zina*)	[§§ 250, 259]

Plates

بسم الله الرحمن الرحيم رب يسر واعن يا كريم
قال الحكيم الرئيس ابو علي بن زين ان النعمة تزرع المحبة ومن
احب المنعم شكره ومن شكره استدام عزه وان يطول بقايه وقد
دعاني صدق المحبة لسيدي ومولاي امير المؤمنين اطال الله بقاه
الى الاجتهاد في شكره وما ادى الى موافقته ومحبته من قول وفعل
واستفراغ الوسع في ذلك واني وجدت في وسعي معرفة اشيا من كتب
علماء الهند والروم وبابل في الطب رجوت ان يكون في انهايها الى امير
المؤمنين اطال الله بقاه قربه اليه وشكر له اذا كانت اسباب السلامة
انفع ما معول عليه ومعمول به في امر الدين والدنيا ولا سبيل الى شي من
ذلك ولا الى امر الاخرة الا بالقوة ولا قوة الا بالصحة ولا صحة الا
باعتدال المزاجات الاربع وقد جعل الله تعالى لها سببا واسبابا
اعلم الخلق بها انبياوه المصطفون صلى الله عليهم ثم الخلفا الذين يرثون
ذلك عنهم وامير المؤمنين ايده الله وارث علم النبوة وجامع خزاين الحكمة
وعند اطبايه من علم الطب والفلسفة ما يقصر عنه علما الهند والروم
مع ما هم عليه من صدق النية ومحض المحبة وشكر النعمة لكن نفسي ابت
الا التقرب اليه ايده الله بما عندي من ذلك بقول مختصر وجيز ليكون
وتذكرة ومثالا له اكرمه الله وللامرا من ولده وولاة عهد المسلمين مد الله

PLATE 1 *Kitāb Ḥifẓ aṣ-ṣiḥḥa* (Incipit, fol. 2b)
MS OXFORD *BODLEIAN* MARSH 413

كتاب اللؤلؤة مما ألفه علي بن رزين في تدبير الصحة ونفي العلة
من البدن باذن الله ودفع الآفة وهو المستعان على ذلك
بسم الله الرحمن الرحيم

PLATE 2 *Kitāb al-Luʾluʾa* (Incipit, fol. 236b)
MS ISTANBUL *AYASOFYA* 3724

Text and Translation

Note on Text and Translation

Generally speaking, the Oxford manuscript (Ox) served as a guide to establishing the text, even though the Istanbul manuscript (Is), in some cases, provided a better reading—either way, the respective variants are documented throughout in the apparatus; dittographies, as they occasionally occur in the Istanbul manuscript, have been emended silently. Certain blurred words or phrases in the second half of the Oxford manuscript, when it becomes the only textual witness, have been restored where possible by consulting relevant parallel passages in the *Firdaus al-ḥikma* (ṬabFir). Bullet points (•) in the apparatus indicate deteriorated letters. The use of bold face and punctuation marks as well as the division of the text into paragraphs are mine.

Greek and Sanskrit source 'verifications', as they appear on occasion in the footnotes to the translation, are by their very nature neither suited nor intended for comparative philological exercises—they merely serve to illustrate the (more or less) *original* shape of Ṭabarī's sources, and also to exemplify the extent of textual alteration (rather than replication); for a discussion of this compositional feature see pp. 23–26 above.

 | DOI:10.1163/9789004445895_003

بسم الله الرحمٰن الرحيم
رب يسّرْ وأعنْ يا كريم

§1 قال الحكيم الرئيس علي بن ربن: إن النعمة تزرع المحبة ومن أحب المنعم شكره ومن شكره استدام عزه وطوّل بقاءه، وقد دعاني صدق المحبة لسيدي ومولاى أمير المؤمنين أطال الله بقاءه إلى الاجتهاد في شكره وما أدى إلى موافقته ومحبته من قول أو فعل واستفراغ الوسع في ذلك، وإني وجدت في وسعي معرفة أشياء من كتب علماء الهند والروم وبابل في الطب رجوت أن يكون في إنهائها إلى أمير المؤمنين أطال الله بقاءه قربة إليه وشكرا له.

§2 إذا كانت أسباب السلامة أنفع مدلول عليه وأفضل معمول به في أمر الدنيا والدين فلا سبيل إلى شيء من ذلك ولا إلى أمر الآخرة إلا بالقوة ولا قوة إلا بالصحة ولا صحة إلا باعتدال المزاجات الأربع، وقد جعل الله تعديلها سبيلا وأسبابا، أعلم الخلق بها أنبياؤه المصطفون صلى الله عليهم ثم الخلفاء الذين يرثون ذلك عنهم وأمير المؤمنين أيده الله وارث عَلَم النبوة وجامع خزائن الحكمة، وعند أطبائه من علم الطب والفلسفة ما يقصر عنه علماء الهند والروم مع ما هم عليه من صدق النية ومحض المحبة وشكر النعمة، لكن نفسي أبَتْ إلا التقرب إليه أيده الله بما عندي من ذلك بقول مختصر وجيز ليكون تذكرة ومثالا له أكرمه الله وللأمراء من ولده ولاة عهد المسلمين مد الله

٢ The opening lines of the Istanbul manuscript run as follows: كتاب اللؤلؤة مما ألفه علي بن ربن
[ربس MS] في تدبير الصحة ونفى العلة من البدن بإذن الله ودفع الداء وهو المستعان على ذلك، بسم الله الرحمٰن
الرحيم ٣ علي : ابو علي Ox || ربن : رَبَّنْ (sic) Ox | قال الحكيم الرئيس علي بن ربن : – Is Ox || شكره[2]
Ox : اشكره Is ٤ طول : ان يطول Ox, احب طول Is ٥ و موافقته : – Is Ox || فعل : عمل Is Ox ٨
إذا : اذ Is Ox || أفضل : – Is Ox || والدين : – Is Ox || فلا : ولا Is Ox ١٠ صلى : صلوت Is Ox ١١
أمير : اسم Is Ox ١٢ الطب والفلسفة : الطسفَة والطب Ox, الملسفه والطب Is || يقصر : فصر Is Ox ||
الروم : الهر Is Ox ١٣ بما : ما Is Ox ١٤ مختصر : فحيص Is Ox || تذكرة : وتذكرة Ox, ىدكوه Is || مثالا
له أكرمه : مىلا اىده Is Ox || للأمراء : الامرا Is Ox

1 The opening lines of the Istanbul manuscript run as follows: "**The Book of the Pearl**, as composed by **ʿAlī ibn Rabban**, on the management of health, the banishment of illness from the body—with the permission of God—and the mastering of malady; this (book) is the aid of choice to achieve (all) that. In the name of God the Merciful the Compassionate".

In the name of God the Merciful the Compassionate.
Lord—pave the way and lend support, O Thou Munificent![1]

The eminent sage **ʿAlī ibn Rabban** says: benevolence fosters love, and he who loves the benefactor is grateful to him, and he who shows gratitude perpetuates his status and prolongs his existence. So it is sincere devotion for my patron and protector, the Commander of the Faithful[2]—may God grant him a long life—, which induces me to engage his recognition, to gain his approval and affection through word and deed, and to do so wholeheartedly. In my pursuit I have acquired (certain) items of knowledge from the medical writings of Indian, Greek and Babylonian[3] scholars which, when brought to the attention of the Commander of the Faithful—may God grant him a long life—, will hopefully draw me closer to him and indicate my appreciation. § 1

If the principles of salvation, in worldly as well as spiritual matters, are the most useful to be pointed out and the most appropriate to be acted upon, then none of this, nor (indeed) the hereafter, can be attained without strength; yet there is no strength without health, and no health without a balance of the four mixtures,[4] whose stability was designed by God as a way and means (to subsist). The most knowledgeable of all mankind in this (respect) are His chosen prophets—may God bless them; then their heirs, the caliphs, (including) the Commander of the Faithful—may God support him—, who inherited the banner of prophethood and who gathers the treasures of wisdom. His physicians understand more about medicine and philosophy than the scholars of India and Greece ever did, even though they (too) had honest intentions, genuine compassion, and gratitude for the privilege. Myself, I seek nothing but his grace—may God support him—when (offering), in a few choice words, what there I have: a reminder and guide (not only) for him—may he be honoured by God—(but also) for the princes among his children, successors to the Muslim throne—may God extend § 2

2 That is the tenth ʿAbbāsid caliph al-Mutawakkil (reg. 232/847–247/861), on whom see KeMut *passim*; cf. also pp. 1, 8 and 26ff. above.

3 On the concept of 'Babylonian' in the present text see p. 26 above.

4 The "four mixtures" (*al-mizāǧāt al-arbaʿ* [sic]) are blood, phlegm, yellow bile and black bile, otherwise called humours (*aḫlāṭ*); cf. also §§ 62–66. For an excellent English summary of the Greco-Arabic physiological system (after al-Maǧūsī, d. late 4th/10th century) see UllIM 56–63.

في أعمارهم ولغيرهم من محيي العلم، فليس كثرة ما في خزائن الخلفاء من أنواع الذخائر والجواهر بمانع التجار من جلب ما عندهم إليهم وعرضه عليهم.

§3 وقد أجمعت العلماء من الأمم كلها على أن الله جعل الصحة سببا من أسباب البقاء والقصد فيها إرفاق له بإذن الله تعالى، وقد أجمعت العلماء على أن الإنسان مخلوق من مزاجات متعادية ويحتاج إلى أغذية وأشربة إن فقدها تلفت نفسه وإن أمعن في الإكثار منها والإقلال أورثته الأسقام والأوهان وإن اقتصد فيها نفعته وقوّتْ جسمه، واتفقت آراؤهم جميعا على أن من جاز الحد في الامتلاء أو الخلاء والسهر أو النوم والحركة أو السكون أو إسهال بطن أو إخراج دم أو إسراف في مباضعة وباه لم يأمن هيجان العلل وبغتات الآفات التي أنا ذاكرها وواصف ما في الاقتصاد من المنفعة وفي السرف والإفراط من المضرة، واتفقوا على أن مَن توَقى ذلك ولزم الاعتدال والقصد رُجيت له الصحة وطول البقاء، ولم أر بين الناس خَلّافا في أن جميع أمور الدنيا من ملك ومال أو لذات أو شهوات إنما هو نبع للبقاء فمن أحب البقاء كرم ما يرفقه ويوافقه وينفعه وهجر في حينه الشهوات ولم يؤثر أكلة على أكلات كما قال بعض الحكماء [من الوافر]:

وَكَمْ مِنْ أُكْلَةٍ مَنَعَتْ أَخَاهَا * بِلَذَّةِ سَاعَةٍ أَكَلَاتِ دَهْرِ
وَكَمْ مِنْ طَالِبٍ يَسْعَى لِشَيْءٍ * وَفِيهِ هَلَاكُهُ لَوْ كَانَ يَدْرِي

---

١ ولغيرهم من محيي العلم Is – : Ox   ٢ عرضه Ox : عرضها Is   ٣ الله + : Ox تبارك وتعالى Is   ٣–٩ من أسباب البقاء ... السرف والإفراط Is – : Ox   ٣ البقاء + : ولا عذبة النافعة Ox ‖ فيها + : سبب من Ox ‖ أسباب البقاء و ... يقوى من Ox   ٧ النوم : النوماء Ox (cf. ܢܘܡܐ)   ٨ باه : باة Ox   ٩ من توفى : من توقى Ox   ١٠ البقاء + : Ox باذن الله Is ‖ خلافا : Ox احلافه Is ‖ ومال Is – : Ox   ١١ لزم : Ox لرمم Is ‖ شهوات Ox : + و Is ‖ نبع : Ox نفع Is ‖ يوافقه وينفعه وهجر في حينه : Ox يسحر فى حينه Is   ١٢ يؤثر Is   ١٢–١٤ كما قال ... يدري Is – : Ox ‖ و + : Ox

---

5 Cf. §§ 27–32 and, by extension, 33–38.

6 These verses, here ascribed to "one of the sages" (*baʿḍ al-ḥukamāʾ*), belong to the poet Ibrāhīm ibn ʿAlī Ibn Harma al-Qurašī (d. c. 176/792), in whose anthology (*dīwān*) they

their lifespans—, and for (all) others who celebrate knowledge. Because the abundance of manifold riches and jewels in the storehouses of the caliphs should not stop the merchants from fetching and offering whatever (goods) they may have!

The scholars of all races are agreed that God established health as one of the pillars of subsistence, and that aspiring to the former is (in itself) a means to achieving the latter—with the permission of God the Sublime. The scholars are (also) agreed that man is composed of antagonistic mixtures, and that he requires food and drink—if he lacks that, he perishes altogether; if he has too much or too little of it, he is left with illnesses and debilities; but if he proceeds with moderation, it will serve him and consolidate his body. All of them (further) share the opinion that no one who exceeds the (proper) boundaries of repletion or depletion, waking or sleeping, movement or rest, purging the belly, bloodletting, or who engages excessively in sexual activities, can be safe from the eruption of diseases and (other) sudden calamities, which I will mention (later on) when recounting the benefits of moderation and the detriments of intemperance and excess.[5] They unanimously state that health and longevity can be hoped for him who adheres to these (rules) and always exercises temperance and moderation. I have never seen anyone denying that all worldly matters—be it property or money, pleasures or desires—are a source of life. Yet he who loves life, values what serves, suits and helps him; he duly resists (excessive) appetites; and he does not devour meal upon meal. In that vein, one of the sages said: §3

How often does a certain bite,
Relished for a brief delight,
Deter the eater's appetite
From all other food for good!

How often does a seeker prize
One thing and this alone,
While in it lies his own demise—
Alas, had he but known![6]

---

take the following form (DīwIH 128): *wa-rubbata aklatin mana'at aḫāhā * bi-laḏḏati sā'atin akalāti dahrī / wa-kam min ṭālibin yas'ā li-amrin * wa-fīhi halākuhū lau kāna yadrī.*

§4 وبلغنا أن عمر بن الخطاب رضى الله عنه قال للحارث بن كلدة: ما الطب يا حارث؟ قال: الأَزْم يا أمير المؤمنين أى الحمية والاقتصاد.

§5 وبلغني أن بعض الفلاسفة كان يحمل على نفسه في الحمية فقال له تلميذه: أيها الحكيم لو زدت في غذائك شيئا ازددت به قوة ونشاطا، قال: أى بُنَيَّ إنما أطلب الغذاء حرصا مني على البقاء ولا أطلب البقاء حرصا مني على الغذاء.

§6 وبلغني أن عمر رأى من ابن له تخمة فقال له: لن تموتَ من هذا لأَصَل عليك، لأنه كان كالمُعِين على قتل نفسه.

§7 ولم أر فيما يستفاد بالمطاعم والمشارب وسائر لذات الدنيا شيئا هو أجل قدرا من الصحة والبقاء، ورأيت مَن أقل من الأغذية والشهوات واقتصر على البلغة والقوت أصح بدنا وأطول عمرا وأقوى شهوات وأخف مؤنات وحركات ممن أكثر منها، وذلك بيّن موجود في أهل البوادي وأصحاب التعب والكد، فهذه محنة صادقة في أن الطب هو الاقتصاد والحمية.

١ بن¹ Ox: اىن Is ‖ للحارث Ox: للحرب Is ‖ ما الطب يا حارث Ox: يا حارث ما الطب Is ‖ الأزم Ox: الآزم Is ٢ أى Ox: ىعنى Is ٣ فقال Ox: فال Is ٤ قال Ox: فمال Is ‖ بنى Ox: + اىى Is ٦–٧ وبلغني … نفسه Ox – :Is ٦ لن تموت: لىن ىتِ Is ‖ عليك: ىلىك Is ٨ شيئا هو Ox: هو شىا Is ٩ رأيت Ox: راىنا Is ١٠ مؤنات و Ox: – Is ١١ التعب Ox: السعد Is

7 That is the second caliph of Islam (reg. 13/634–23/644), on whom see LB'Um *passim*.

8 al-Ḥāriṯ ibn Kalada aṯ-Ṯaqafī, a contemporary of the prophet Muḥammad, is one of the earliest *Arab* physicians known to us and no doubt a historical person, even though later biographical accounts are somewhat incoherent and embellished with legendary stuff; al-Ḥāriṯ died sometime between the late 630s and early 660s CE, depending on which account one wishes to rely—see UllMed 19 f. and GaS 3/203 f. (with the sources quoted there); cf. also the recent appraisal InkḤā *passim*.

9 A parallel version of this anecdote is given IAU 2.1/304,8 f., referring to the Islamic tradition (*ḥadīṯ*) and running as follows: "'Umar asked al-Ḥāriṯ ibn Kalada: 'What is medical treatment (*diwā'*)?' He replied: 'Restraint, meaning diet (*ḥimya*)'". A variant transmission is found in the same source (2.1/304,5 f. after Ibn Ǧulǧul [IĞṬab 54,5 f.]), featuring, instead of 'Umar, the first Umaiyad caliph Mu'āwiya and running as follows: "Mu'āwiya said to him [scil. al-Ḥāriṯ]: 'What is medicine (*ṭibb*), O Ḥāriṯ?' He replied: 'Restraint, meaning hunger (*ǧū'*)'"; the latter variation is also recorded ZauMuḫ 162,19 f. It may be worth noting that the term *azm*, here translated "restraint", literally means "a single act of eating, i.e. an eat-

We were told that ʿUmar ibn al-Ḫaṭṭāb[7]—may God be pleased with him—said to al-Ḥāriṯ ibn Kalada:[8] "What is medicine, O Ḥāriṯ?" He replied: "Restraint, O Commander of the Faithful, namely diet and moderation".[9] §4

I was told that one of the philosophers used to subject himself to dieting. So his pupil said to him: "O Master, if you took a bit more food, you would gain strength and energy!" He replied: "O Sonny, I seek food because I am keen to stay alive; I do not seek to stay alive because I am keen on food".[10] §5

I was told that ʿUmar,[11] seeing one of his sons suffer from (gluttonous) indigestion, said to him: "You will not die from this bad habit!"—because he behaved like someone who is assisting his own death.[12] §6

Amidst (all) the benefits that may be derived from food and drink and other pleasures of this world, I have found nothing of greater value than health and survival. I have (also) found that he who shows restraint in nourishments and desires, and who confines himself to the vital necessities, has a healthier body, a longer lifespan, more vigour, (needs) less provisions and is more agile than someone who overdoes it—which can clearly be seen in desert dwellers and heavy labourers. This is a sound proof that medicine (involves) moderation and diet. §7

---

ing but once in the course of the day", see LaLex 1/55a; cf. also ibid. 1/54c, where the same term, in the context of the same anecdote, is explained by "the practising [of] abstinence […] or the not putting in food upon food".

10 This little story is another variation on a theme found in almost every Greek and many Arabic gnomologia; the philosopher in question is generally Socrates (rarely Diogenes). The core message (i.e. eat in order to live rather than live in order to eat) may be placed into a different narrative frame, for example: "A philosopher wrote to him [scil. Socrates], reproving him for eating little and wearing coarse clothes, saying: 'You claim that compassion (*raḥma*) is the duty of anyone endowed with spirit and soul. Now, you (too) are endowed with spirit and soul, and yet you make them both suffer by diminishing their food and by clothing them coarsely!' (Socrates) replied to him: 'You have upset me (*mauwaǧtanī*) to my face, which is a shame. You have reproved me for wearing coarse stuff, but sometimes a man loves an ugly woman and leaves a beautiful one; and you have reproved me for eating little, but I only eat in order to live whilst you live in order to eat. Greetings'". See, for this version of the story, GuWL 86–89 (no. 8a) and, for further sources, ibid. 287.

11 Cf. note 7 above.

12 ʿUmar had several sons, see e.g. ISBio 190. The anecdote itself is, as far as I can tell, not recorded elsewhere.

§8 واجتمعت آراء العلماء على أن الطب هو حفظ الصحة ونفى العلة بإذن الله تعالى ودفع الداء بضده، وذلك كمن يعتل من الامتلاء فينفعه الخلاء وكمن يعتل من الخلاء فينفعه الاغتذاء أو كمن يهيج به الحرارة فينفعه التبريد بالأطعمة والأشربة والأدهان الباردة أو يعتريه البرودة فينفعه التسخين بالأطعمة والأشربة والأدهان الحارة أو يغلب عليه يبس فينفعه التليين أو يسترخي بدنه من فرط التليين فينفعه التدبير المتيبس أو كمن يمرض من شدة التعب والكد فتنفعه الراحة، فهذا هو الطب وإليه يقصد الأطباء في علاجاتهم.

§9 وقد يشتمل كتابي هذا على أبواب من عيون الطب بينة سهلة ليس فيها تلبيس ولا غويص، مَن عرفها عاين الطب عيانا ووجد عليه برهانا بوصف مقنع يسير غير عسير.

في حفظ الصحة

§10 قال جالينوس: إن حفظ الصحة يكون بإذن الله تعالى على وجهين، أحدهما الاغتذاء بما يوافق سن الإنسان وزمان السنة التي هو فيها والعادة التي اعتادها والأطعمة والأشربة التي ألفها ونبت بدنُه عليها، والوجه الثاني إخراج ما يتولد فيه من الفضول والمواد الرديئة.

١ آراء Ox: –Is ‖ على Ox: –Is ‖ تعالى Ox: –Is ٢ بضده Is: بضدة Ox ‖ من الامتلاء Ox: بالامـلا
فينفعه الاغتذاء أو Is ‖ فينفعه Ox: ونـمعه الاعتدال و Is ٣–٤ والأدهان ... الباردة Ox: – Is ٤ فينفعه
Ox: فمـمعه Is ٥ التدبير المتيبس: التدبير التيـيس Ox, الـلـس الملـس Is ‖ التعب Ox: المرض
Is ٧ يشتمل Ox: بدل Is ‖ عيون Ox: عـوب Is ‖ بينة سهلة Ox: مسهله Is ٨ بوصف مقنع يسير غير
عسير Ox: – Is ١٠ إن Ox: – Is ‖ يكون Ox: تكون Is ‖ بما Ox: به انما Is ١١ فيها Ox: فه Is

13 Typically Islamic eulogies and pious evocations, when they occur in passages attributed to *non-Arabic* sources, are obviously later additions to the original texts; some such phrases may already have formed part of the lost autograph of the work, but they may equally, and perhaps more likely, trace back to subsequent copyists.

Scholars concur in the view that medicine (means) preserving health, averting sickness—with the permission of God the Sublime—, and combating a disease with its opposite. For example, like someone who suffers from repletion is helped by depletion, whilst someone who suffers from depletion is helped by alimentation; or like someone who is assailed by heat is helped by cooling-off through cold foods, drinks and oils, whilst someone who is overcome by coldness is helped by heating-up through hot foods, drinks and oils; or (like) someone who is seized by dryness is helped by laxation, whilst someone whose body has become slack due to overmuch laxation is helped by a drying regimen; or like someone who falls ill in the toils of hard labour is helped by rest. This, then, is medicine and what physicians aim to achieve with their treatments. § 8

My present book comprises (several) chapters, (drawn) from the fountains of medicine, (written in) clear and simple (language), with no pretence nor profundity; he who peruses these (chapters) will see medicine at first hand and find its validity confirmed in an adequate, easy and uncomplicated style. § 9

On the Preservation of Health

Galen said: health is preserved—with the permission of God the Sublime[13]—in two ways. First, by a diet that is appropriate to a person's age, to the time of year in which he finds himself, to the habits he has grown into, and (that allows for such) foods and drinks which he likes and on which his body thrives; the second way is by dislodging the residues and bad matters which are generated inside of him.[14] § 10

14 This appears to be an amplification of the following Galenic statement: "Quum una fit ars, quae circa corpus hominis occupatur, […] ejus primae ac maximae partes sunt duae; quarum altera sanitatis conservatrix, altera curatrix appellatur", see GalKü 6/1 = GalKo 3 (*Ὑγιεινῶν λόγοι*, cf. FiCG no. 37). See moreover GalKü 6/7 = GalKo 5f. (op.cit.): "Quum enim quotidie propter insitum ipsis calorem cunctis animalibus multa substantiae portio defluat, indigeamus autem ad hujus commoderationem conservandam cibo, potu, respiratione et pulsu, necesse est ex iis sequi excrementorum proventum"; further, GalKü 6/8 = GalKo 6 (op.cit.): "Quum igitur edere bibereque necessaria animantibus sint, haec autem subsequatur excrementorum generatio"; and GalKü 6/9 = GalKo 6 (op.cit.): "Atque tibi geminum jam scopum salubris victus sermo noster exposuit; alterum vacuatorum repletionem, alterum excrementorum excretionem".

§11 وقال أيضا: إنه لما كانت أبدان الناس وما يصل إليها من الأغذية تحلل وتنفش أولا فأولا بالحرارة الغريزية وبحرارة الشمس والرياح التي تنشف الرطوبات من الأبدان كلها ومن الأنهار والبحار أيضا احتاج الإنسان لذلك إلى أن يغتذي بما يقوم مقام ما يحلل وينفش من بدنه أولا فأولا، فإذا كان البدن متخلخلا حارا سخيفا نفعته الأطعمة الغليظة لأن ما ينفش ويحلل من مثل ذلك البدن يكون كثيرا لِسعة منافذه وكثرة حرارته وما كان من الأبدان ملززا مكتنزا يابسا فإنه تنفعه الأشياء الرطبة اللطيفة لأن الذي يحلل من مثل هذه الأبدان يكون قليلا لضيق منافذه.

§12 وقال أيضا: إن الوجه في حفظ الصحة بإذن الله تعالى أن يغتذي الرجل بما يوافق مزاج بدنه في حال صحته فمن كان في حال صحته حار المزاج وافقته الأشياء الحارة المعتدلة ومن كان بارد المزاج وافقته الأشياء الباردة المعتدلة وكذلك القول في الرطب واليابس من المزاجات، فإن زادت الحرارة والتهبتْ التهابا انتفع حينئذ بما يضادها ويخالفها من البرودات وإن هاجت البرودات انتفع حينئذ بما يضادها من الحرارات، وإذا كانت المعدة حارة قوية جدا كان أنفع الأغذية لصاحبها ما غلظ وقوى كالنار الغليظة التي تقوى على إحراق وإكلال الجزل من الحطب وإذا كانت باردة ضعيفة كان أنفع الأغذية له ما خف واستمرأ كالنار الضعيفة التي توقد بالقصب ودقاق الحطب.

§13 ومن الدليل على الاستمراء خفة البدن وصفو الجشاء وحركة الشهوة، فأما الدليل على التخمة فاسترخاء البدن والكسل وانتفاخ الوجه وكثرة التمطي وثقل العين وأن يصير الجشاء منتنا أو

---

١ من الأغذية Ox : الاعذيه والاسربه Is   ١–٢ أولا فأولا بالحرارة الغريزية Ox : وبعدى منها Is   ٢ الرياح Ox : ارواح Is   ٣ إلى Is : – Ox ‖ مقام Ox : – Is ‖ فأولا Ox : اولا Is   ٤ متخلخلا Ox : متحللا Is ‖ سخيفا Ox : سحقا Is ‖ ينفش Ox : ينفس Is ‖ مثل Is : – Ox   ٦ هذه الأبدان Ox : البدن Is ‖ لضيق منافذه Ox : – Is   ٧ قال Ox : فالوا Is ‖ حفظ Is : – Ox ‖ تعالى Ox : – Is   ٨ حال[2] Is : – Ox ‖ الحارة Ox : الباردة Is ‖ من Ox : ان Is   ٩ الباردة Ox : الحاره Is   ١٠–١١ وإن هاجت ... الحرارات Ox : – Is   ١٢ وإذا Ox : فاذا Is   ١٢–١٣ كالنار ... الغليظة Is : – Ox   ١٣ له Ox : لما Is   ١٤ الدليل[2] Ox : – Is   ١٥ الكسل Ox : + واسماخ البدن والكسل Is

---

15 The "innate heat" (*al-ḥarāra al-ġarīzīya* < *τὸ ἔμφυτον θερμόν*) is the central exponent and custodian of all animal life; it is generated and stored in the heart; it maintains the vital functions of the organism, but at the same time contributes to its decline by using up 'fuel' that becomes ever more rare with senescence; its ultimate exhaustion results in

He also said: human bodies, together with any nourishment that enters them, gradually dissolve and subside under (the influence of) the innate heat,[15] the warmth of the sun, and the winds which invariably dry out bodily moistures just like (they dry out) rivers and lakes;[16] therefore, man needs to nourish himself (in order to) replace what his body loses through gradual dissolution and subsidence. A body that is hot, fragile and unstable benefits from tough food, because such a body, due to the wideness of its pores and the profusion of its heat, loses a lot; whereas bodies that are dry, firm and stable benefit from moist and soft stuff, because such bodies, due to the narrowness of their pores, lose less. § 11

He also said: the guiding principle for preserving health—with the permission of God the Sublime—is that a man eats what suits the mixture of his body in a state of health.[17] For him whose mixture in a state of health is hot, moderately hot stuff is suitable; for him whose mixture is cold, moderately cold stuff is suitable; and the same rule (applies) to moist and dry mixtures. When heat increases and flares up, he benefits from cold stuff which opposes and counters it; (equally) when coldness rises, he benefits from opposing hot stuff. He whose stomach is hot and really strong, benefits most from tough, consolidating food, just like a forceful fire which is able to wear down and burn up (even) thick chunks of wood; yet he whose (stomach) is cold and weak, benefits most from light, easily digestible food, just like a feeble fire which is kept burning with reeds and wood shavings. § 12

Indications of a (good) digestion are lightness of the body, pure belching, and a brisk appetite; as for indications of a bad digestion, (these are) laxity of the body, sluggishness, a puffy face, frequent stretching, heavy eyes, belching § 13

---

death. On this core principle of humoral physiology see UllIM 65ff. and, for a wider contextual frame, MenHL 8–26 and *passim*.

16 GalKü 10/753f. (*Θεραπευτικὴ μέθοδος*, cf. FiCG no. 69): "Quaenam igitur est putredinis natura? Mutatio totius putrescentis corporis substantiae ad corruptelam ab externo calore. Non enim profecto a proprio calore corrumpitur quicquam, imo vero contra quodcunque et augetur et roboratur et sanum est et vivit eorum quae sunt quodque proprio calore regitur. […] His vero laesis una etiam naturalis calor laeditur; alter vero quidam alienus, ac praeter naturam in corporibus excitatus, primum quidem ipsos humores propter humiditatem et putrefacit et corrumpit, spatio vero temporis tum adipem invadit tum etiam carnem".

17 GalKü 1/655 = GalHe 91 (*Περὶ κράσεων*, cf. FiCG no. 9): "Omne animal conveniente sibi nutritur alimento, conveniens autem cuique alimentum est, quicquid assimilari corpori, quod nutritur, potest".

حامضا وتهيج قراقرَ ونفخا وفتورا وتجلب البلل في الفم والمُطَواء والثُؤَباء والعُرَواء أعني بالمطواء التمطي وبالثؤباء التثاؤب وبالعرواء القُشَعْريرة، وما يعين على الشهوة للطعام النظر إلى ألوانه ووصفها وشم أرايحها والنظر إلى من يجيد أكلها واستعمال المشي والركوب وأنواع الضمور قبل الأكل ودخول الحمام والاستياك وأخذ الهليلج المربى وجوارش من الجوارشات.

§14 وقال جالينوس: ينبغي للرجل إذا انتبه من نومه أن يتمشى قليلا ويغمر رقبته ورأسه نعما ويتمرخ بدهن موافق لسنّه وزمانه وأن يمتشط فإن الغمر المعتدل يصلب البدن والتدهين يلين الجلد والمشط يخرج البخارات من الرأس، فأما الغذاء المفرط الدائم فإنه يرخي البدن ويورث الفتور فإذا اشتهى أكل ما يوافق طبيعته وزمانه فلا يأكل في الشتاء باردا ولا في الصيف حارا سخنا، فإذا نام فلينم على يساره ساعة ثم ينقلب على يمينه لأن الشق الأيسر بارد فهو يحتاج إلى ما يسخنه، ويبدأ أولا بما لان من الغذاء والثمار وما كان منها مريئا سهلا ثم يأكل بعده ما صلب منها لأن الطعام اللين المريء يخرج من البدن سريعا فيسهل لذلك خروج الطعام الصلب بعده فإن هو بدأ بالطعام الصلب البطيء الهضم ثم أكل بعده ما لان من الطعام والثمار انهضمت تلك الأطعمة اللينة سريعا وطلبت مخرجا واحتبست تلك الغليظة تحتها غير منهضمة فإذا لم تجد تلك الخفيفة مخرجا فسدت وأفسدت ما تحتها من الأطعمة الغليظة بفسادها.

---

١ تهيج قراقر : يهيح قراقراً ,Ox يهسح قراقر Is || نفخا وفتورا : Ox نفح وفسور سهوه Is || تجلب : يجلب
Ox || ٢ بالثؤباء : الثوبا Ox ١ العرواء : العَزْوَاء Ox ١–٢ والمطواء ... القشعريرة : – Is بحلل ,Ox Is
بالعرواء القشعريرة : بالعَزوا الاقشعريره Ox || ما : مما Is || للطعام : فى الطعام Is ٣ أرايحها : Ox
ارباحها Is || إلى : – Is || الضمور : الصموت Is ٤ الاستياك : الاسهال Is || الهليلج Ox
الهلىح Is || جوارش من الجوارشات : حوارسن من الحوارسنات Is ٥ من نومه : – Is ٦
أن : – Is || الغمر : الغمز Is ٧ من الرأس : – Is || الغذاء : للغمر Is ٨ فلا : ولا Ox
في الشتاء باردا : بالستا بارد Is || فإذا : واذا Is ٩ يبدأ : سداوا Is ١٠ منها[2] : – Ox Is
اللين المريء : المرى اللىں Is ١٢ الهضم : – Is || والثمار : – Is ١٢–١٣ وطلبت ١٠–١١ Is
مخرجا واحتبست : فطلىت المخرح وىحىس Is

---

18 It is not clear why Ṭabarī would have felt the need to explain the synonymous, genuinely Arabic terms *muṭawāʾ* by *tamaṭṭin* (shared root *mṭw*), *ṯuʿabāʾ* by *taṯāʾub* (shared root *ṯʾb*) and *ʿurawāʾ* by *qušaʿrīra*, nor indeed why he did not settle on single-term definitions in the first place.

that turns foul or sour, the onset of (gastric) rumbling, bloating and flagging, an accumulation of fluids in the mouth, as well as *muṭawāʾ*, by which I mean craning, *ṯuʾabāʾ*, by which I mean yawning, and *ʿurawāʾ*, by which I mean quivering.[18] What promotes an appetite for food is looking at its different kinds, (hearing) them described, smelling their fragrances; watching somebody who eats heartily; walking, or riding, or doing (other) kinds of exercise before a meal; visiting the bathhouse; (minding) dental hygiene; and taking (some) preserved myrobalans, or one of the stomachics.[19]

And Galen said: when he wakes up from his sleep, a man is well advised to stretch his legs a little, to submerge his head and his neck fully (in water), to rub his skin with an oil that is appropriate for his age and the time (of year), and then to comb his hair—controlled submersion toughens the body, embrocation softens the skin, and combing releases vapours from the head.[20] As constant overeating slackens the body and bequeaths fatigue, he should, when he feels an appetite, eat what is appropriate for his constitution and the time (of year), avoiding cold (stuff) in winter and warm or hot (stuff) in summer. When he goes to sleep, he should lie on his left side for a while, then turn over on his right side, because the left side is cold and needs warming up. (When eating), he should start with soft food and fruits, with what is easily digestible, and only then reach for something hard—(this is so) because soft, digestible food exits the body quickly and thereby facilitates the passage of hard food. For if he starts with eating hard, slowly digestible food, and then follows it with soft stuff and fruits, the latter, digested faster, seek an exit but are held back behind that tough, undigested (food); and if the light (stuff) cannot find an exit, it becomes corrupted, and through its corruption corrupts the tough food below it. §14

19 The term *ǧawāriš* "stomachic" (pl. *ǧawārišāt* < Persian *guwārišn* "medicamentum compositum, quod cibi digerendi caussa edunt" [VuLex 2/1040a–b]) was used to denote a particular pharmacological category, as can be seen already in the dispensatory of Ṭabarī's contemporary Sābūr ibn Sahl (d. 255/869) who dedicated a whole chapter specifically to the preparation of 'stomachics', cf. SābAq 137–155 = SābDis 116–131.

20 These recommendations regarding early morning regimen cannot be rescued in one piece from Galen's extant writings.

§15 وقال بقراط المتقدم: إن مما يعين على الصحة أن لا يأكل الرجل حتى يتعب قليلا ثم يستريح إلى أن يشتهى الطعام فإذا أكل اضطجع ونام، فإن أحس بثقل في شراسيفه نفعه أن يضع على بطنه مرفقة أو ينام عليها أو على البطن أو يعانق صبية حارة أو شيئا حارا أو يجعل وسادَه مرتفعا، فإن تجشأ جشاء حامضا دل ذلك على برد المعدة فليشرب الماء الحار بالسكنجبين ثم يتقيأ، فإن أحس بثقل في كبده فليشرب السكنجبين أو الجوارش الكموني، وإن عطش ليلا ثم لم يشرب كان أقوى لحرارته، فإن اعتياد شرب الماء ليلا والإكثار من مياه الجليد والثلج مما يطفئ الحرارة الغريزية ويورث الأسقام والسل إلا أن يكون ذلك العطش من حمى ملتهبة أو من مطاعم ومشارب حارة جدا أو مالحة.

§16 وقال: إن الاستحمام قبل الطعام نافع لأنه يذيب الفضول ويخرجها بالعرق، فأما بعد الطعام فرديء لأنه يورث سدد الكبد.

§17 وقال أبقراط أيضا: إن الحركة والضمور قبل الطعام يوقدان نار المعدة، فأما بعده فرديء لأنه ينزل الطعام غير نضيج فيورث ذلك سددا وأسقاما.

١ المتقدم إن مما Ox : الحكـم المعـن انما Is ٢ يشتهى Ox : سهى Is ٣ أو[5] Ox : و Is || وساده Is : وسادةً Ox ٤ بالسكنجبين Is : بالسكنجيين Ox ٤–٥ السكنجبين ... ثم يتقيأ Ox : – Is ٥ السكنجبين Ox : السكنجيين Is || الجوارش Ox : الحوارسن Is ٦ مما Ox : فما Is ٧ ملتهبة Ox : ملهبه Is ٨ أو مالحة Ox : – Is ١١ أبقراط أيضا Ox : بقراط Is ١٢ فيورث Ox : فورّث Is

21 HippLi 5/314 f. (*Ἐπιδημίαι*, cf. FiCH no. 19): “Que les exercices précèdent les aliments”.

22 For an early example of the so-called *cuminy* stomachic (*al-ǧawāriš al-kammūnī*) see SābAq = SābDis no. 217.

23 HippLi 2/368 f. = HippKü 143 (*Περὶ διαίτης ὀξέων*, cf. FiCH no. 4): “Le malade ne doit pas se baigner quand il vient de prendre de la *ptisane* [barley porridge] ou quelque boisson; il ne doit, non plus, prendre ni *ptisane* ni boisson immédiatement après être sorti du bain”.

Hippocrates the ancient said: conducive to health is that a man only eats after having done a bit of exercise and then taken a rest, until he feels an appetite for food;[21] when he has eaten, he lies down on his side and sleeps. If he (then) has a sensation of heaviness in (the region of) his costal arches, he is helped by putting a pillow on his belly, or by sleeping on it, or (by sleeping) on the belly, or by embracing a sultry girl or some (other) hot thing, or by resting on his bed in an elevated position. If he is troubled by sour belching, it means that his stomach is cold, (in which case) he drinks warm water with oxymel, then vomits. If he has a sensation of heaviness in his liver, he drinks oxymel or the *cuminy* stomachic.[22] If he is thirsty in the night and does not drink, he will get hotter; (however), the habitual consumption of water at night, as well as overmuch ice water or snow water, extinguish the innate heat and leave consumption and (other) illnesses, except when this thirst is due to a flaming fever, or caused by very hot or salty foods and drinks. §15

He said: bathing before a meal is useful, because it melts the residues and drives them out through sweat; yet it is bad after a meal, because it brings obstructions of the liver.[23] §16

Hippocrates also said: movement and exercise before a meal ignite the fire of the stomach; yet they are bad afterwards, because (then) the food descends in an uncooked (state), which brings obstructions and (other) illnesses.[24] §17

24 HippLi 2/326–329 = HippKü 133 (*Περὶ διαίτης ὀξέων*, cf. FiCH no. 4): "Si, dans le passage d'une alimentation abondante à l'abstinence, il faut donner du repos au corps, il faut aussi, quand on fait succéder subitement le repos et l'indolence à une grande activité corporelle, donner du repos au ventre, c'est-à-dire diminuer la quantité des aliments; sinon il en résultera, pour tout le corps, de la souffrance et une pesanteur générale"; cf. also note 21 above.

§18 وقال: النوم قبل الطعام يهزل البدن وينشف رطوبته، والنوم بعد الطعام يغذو ويقوي لأنه إذا نام الإنسان برد ظاهر البدن واجتمعت الحرارة الغريزية المنتشرة في البدن كله إلى المعدة وما والاها فتقوى حينئذ المعدة على الإنضاج.

§19 ومدح أبقراط الطعام عِشاءً فقال: إن من اعتاد العَشاء ثم تركه يبس عليه طبيعته، وذكر أن الغذاء يستقبل حر النهار مع شغل الحواس والنفس بما يسمعه الإنسان ويباشره ويفكر فيه فتنتشر الحرارة الغريزية في ظاهر البدن فتضعف المعدة لذلك عن إنضاج الطعام، فأما العشاء فإنه خلاف ذلك لأنه يستقبل سكون البدن وهدوء الحواس والنفس وهجوم الليل البارد الذي تهرب الحرارة الغريزية منه إلى غور البدن وقعره.

في الأغذية والتدابير التي تقوّي البدن وتسمّن وتهزّل وتهيّج وتسكّن

§20 إني رأيت من الأشياء أشياء تقوي البدن وأشياء توهنه وأشياء تيبسه وأشياء تنشطه وأشياء تهزله وأشياء تسمنه وأشياء ترطبه وتهيجه على الباه وأشياء تورثه الملالة والفتور.

§21 فمما يقويه الأغذية والأشربة الخفيفة الموافقة إذا تناولها الإنسان في أوقات الحاجة إليها فإن أكثر منها أورثته التخم والسدد وأنواع الأسقام مع استرخاء البدن وتتابع التثاؤب والتمطي والقراقر

١ النوم[2] Ox : هو Is || لأنه Ox : لان Is ٢ المنتشرة Ox : المسفسه Is || المعدة Ox : المعمده Is ٤
أبقراط الطعام عشاء Ox : بقراط العسا Is || اعتاد العشاء Ox : اعتاده Is ٥ يستقبل Ox : يسعمل
Is || يسمعه Ox : يسمع Is || فتنتشر Ox : وبس Is ٦ عن إنضاج الطعام Ox : انضاح البدن Is ٧
هدوء Is : هذوّ Ox || الذي : التى Ox, السى Is ٨ قعره Ox : المعدة Is ١٠ وتسكن Ox : على الباه Is ١١
أشياء[1] Ox : ما Is || البدن Ox : بدن الانسان Is || وأشياء تيبسه وأشياء تنشطه Ox : – Is ١١–١٢ أشياء
تهزله وأشياء تسمنه Ox : اسا بسمنه واسا بهرله Is ١٢ ترطبه Ox : + وبسطه Is ١٣ فمما Ox : فاما
ما Is || الأغذية Ox : فالاغديه Is || الخفيفة Ox : – Is ١٤ استرخاء Ox : الاسترخا Is

25 HippLi 6/572–575 = HippJo 182 ff. (*Περὶ διαίτης*, cf. FiCH no. 36): “Le sommeil, à jeun, atténue et refroidit, à moins qu’ il ne soit prolongé, évacuant l’humide qui existe; s’ il est

And he said: sleeping before a meal emaciates the body and dries up its moisture; sleeping after a meal nourishes and strengthens, because when a person sleeps, the surface of the body cools down and the innate heat, which is (otherwise) spread all over the body, gathers in the stomach and adjacent areas, thus empowering it to thorough cooking.[25] §18

Hippocrates approved of eating in the evening—he said: somebody who is used to having a meal in the evening, then quits (that habit), such a one's nature becomes dry. He (further) mentioned that meals (taken earlier) are met not only by the heat of the day but also by a preoccupation of the person's soul and senses with what he hears, attends to, and thinks about; then the innate heat spreads out across the surface of the body, which (in turn) disempowers the stomach from thoroughly cooking the food. An evening meal is different in that it is met by a body at rest, a stillness of soul and senses, and the fall of a cool night, from which the innate heat flees deep into the depth of the body.[26] §19

On nourishments and (other) measures
which strengthen, fatten, thin, arouse or appease the body

I look at the things (and see that) some strengthen the body, some weaken it, some desiccate it, some stimulate it, some make it thin, some make it fat, some moisten it and arouse (a desire) for sexual intercourse, and some leave it with languor and lassitude. §20

Among (the things) that strengthen the body are light, suitable foods and drinks, if a person reaches for them as needed; if he overdoes it, they cause indigestion, obstructions, (all) kinds of laxity-related illnesses, serial yawn- §21

---

prolongé davantage, il échauffe, il fond la chair, il résout le corps et l'affaiblit. Après le repas, il échauffe et humecte, répandant la nourriture dans le corps. C'est surtout après les promenades du matin que le sommeil dessèche. Les veilles sont nuisibles après le repas, ne permettant pas à l'aliment de se fondre; à jeun, elles produisent, il est vrai, une certaine atténuation, mais elles sont moins nuisibles. L'inaction humecte le corps et l'affaiblit; car l'âme, demeurant immobile, ne dissipe pas le liquide du corps".

26 The *psychological* ideas insinuated in this paragraph are not of Hippocratic (nor indeed Galenic) origin; as for some of the other statements made here by Ṭabarī, there are only indistinct correspondences between them and the passages HippLi 1/590–595 = HippHe[1] 42f. (*Περὶ ἀρχαίης ἰητρικῆς*, cf. FiCH no. 1) and HippLi 2/280–297 = HippKü 122–125 (*Περὶ διαίτης ὀξέων*, cf. FiCH no. 4), largely dealing with changes in eating habits.

والنفخ والخدر، فأما ما يسمنه ويرطب بدنه فالراحة والدعة وأكل الإسفيدباجات والأطعمة الحارة الرطبة والاقتصاد في ذلك كله وشرب الأشربة المعتدلة في حرها ويبسها الممزوجة الرطبة والنوم بعد الطعام على الفُرُش والحشايا الوثيرة اللينة فإن الحرارة الغريزية تجتمع عند ذلك لإنضاج الغذاء وأن يستعمل الباه على قدر القوة والاستحمام بالمياه الحارة العذبة وقلة اللبث في الحمام لئلا يأخذ الحمام من رطوبته وبلته بل يأخذ البدن من بلة الحمام ورطوبته ويشم الرياحين والأفاويه المعتدلة كالنرجس والسوسن والخيري وما وافق البدن في كل زمان مثل الياسمين في الشتاء والورد والبنفسج في الصيف وأن يتعاهد القيء أحيانا ولا سيما في الصيف فإن الرطوبات تطفو فيه على المعدة ولأن في القيء غسلا وتنقية للمعدة من الرطوبات والمواد الرديئة فإذا قلت تلك المواد فيها قويت الحرارة الغريزية على هضم الأغذية وابتل البدن لذلك وامتلأ وأنفع من ذلك كله للترطيب والسمنة فيما رأيت الغَناء والفرح والغلبة للأعداء ودرك الرجاء، فأما ما يهزل البدن ويجففه فخلاف ذلك كله من قلة المطعم والمشرب وشدة الغضب والحركات في الشموس والسمائم والسهر الطويل الدائم والنوم قبل الطعام على الفُرُش الخشنة وعلى التراب والرمل لأن الحرارة تقبل حينئذ على ما في البدن من رطوبة فتنشفها وتحيف عليها وتنال منها والاستحمام بالمياه الكبريتية والمالحة وأكل الأطعمة الحارة اليابسة والحامضة والمالحة والحِرِّيفة والقلايا المعمولة بالأبازير الحارة الجافة وشرب الأشربة صرفا والإكثار من إسهال البطن أو إخراج الدم أو الإفراط في المجامعة فإن ذلك كله يهزل البدن ويذيب دسمه ورطوباته وأقوى من ذلك كله على إنهاك البدن وإذابته فيما رأيت الفقر والخوف والفِكَر الرديئة والهموم.

---

١ النفخ : + والسدد Ox, السدد Is || فأما : Ox واما Is || الإسفيدباجات : Is الاسفيفدفاجات Ox ٢ الحارة : الجارة Ox, و الحلوه Is ٣ والحشايا الوثيرة : Ox الوسـره Is || لإنضاج : Is الانضاج Ox ٥ بلته : Is يلتذ Ox ٦ الخيري : Ox الحـرـد Is || مثل : Ox من Is || في الشتاء : Ox – Is ٧ البنفسج : Ox السـفـح Is || يتعاهد : Ox بعـاد Is || فيه على : Ox على راس Is ٨ للمعدة : Ox المعده Is || فإذا : Ox فـاذ ما Is ٩ لذلك : Ox – Is || للترطيب : Ox لترطيب الـدن Is ١٠ للأعداء : Ox عن الاعدا Is || فأما : Ox واما Is || يجففه : Ox بيبسه Is ١١ الغضب : Ox الـنعـب Is || الشموس : Ox السموم Is ١٢ السهر : Ox السعز Is || الفرش : الفُرشه Ox, الشى Is || وعلى التراب والرمل : Ox – Is ١٣ وتحيف عليها : Ox – Is ١٤ اليابسة : Ox + والـبـاسه الـبـارده Is || الحريفة : Is الحريفيه Ox ١٥ الجافة : Ox الحارفه Is || الأشربة : Ox الادويه Is || أو¹ : Ox و Is ١٥–١٦ أو الإفراط في : Ox وفرط Is ١٦ يذيب : Ox يذوب Is ١٧ والهموم : Ox – Is

ing, stretching, rumbling, bloating, and numbness. As for those (things) that fatten and moisten the body, (they consist in) rest and composure; eating thick bouillons and hot-moist dishes, all within reason; drinking moderately hot beverages which are mixed from dry-moist (ingredients); sleeping after a meal on soft, comfortable mattresses and cushions, for then the innate heat gathers to a thorough cooking of the food; practising sexual intercourse to the extent of one's strength; bathing with warm, fresh water without lingering in the bathhouse, such that the bath does not take any moisture or fluid from the body but rather that the body takes moisture and fluid from the bath; smelling aromatic plants and mild spices, like daffodils, lilies, gillyflowers, and anything that is seasonably suitable for the body (to smell), for example jasmines in winter and roses or sweet violets in summer; making sure to vomit from time to time, especially in summer, because then dampnesses flood the stomach, and (also) because vomiting (generally) washes out and cleanses the stomach of bad fluids and matters—when that stuff decreases in the stomach, the ability of the innate heat to digest food increases, which (in turn) hydrates the body and fills it out; yet more useful than all of this for gaining moisture and corpulence is, as I have seen, living in comfort, being joyful, defeating one's enemies, and fulfilling one's hopes. As for those (things) that make the body thin and dry it out, (they consist in) the opposite of it all, namely taking little food and drink; being very angry; moving about in the sun and in hot winds; having chronic insomnia; and sleeping before a meal on coarse mattresses or on soil and sand—because then the (external) heat turns upon the (internal) moisture and devours, diminishes and destroys it; (further), bathing with sulphurous or saline water; eating hot-dry, sour, salty or pungent foods, or pan-fried meals made with hot, dried seasonings; drinking undiluted wines; overmuch purging of the belly or bloodletting; and excessive sexual intercourse, for any of that emaciates the body and melts away its fat and moistures; yet stronger than all of this in terms of exhausting and wasting the body is, as I have seen, living in poverty, being fearful, having bad thoughts, and worrying.

§22 وما يفرح القلب ويهيج على الباه تعهد البدن بالطهارة والاستحمام بالمياه الحارة ووجدان الشهوة للمألوفة المعتادة من الأطعمة والأشربة والرياحين والطيب ولزومها والملابس المصبغة والأشياء التي تفرح وتعجب بها النفس ويدخلها لها الابتهاج والعجب والشرب مع الأحبة والنظر إلى الوجوه الرائقة المعشوقة وتعهد الاستياك والاكتحال والإدهان بأدهان موافقة للبدن والنظر إلى الحيوانات إذا سافدت والفكر في أنواع الجماع والنظر في الشِعر والكتب التي تصف ذلك وتحكيه واستماع الأغاني والملاهي التي تشوّق إلى الحلائل والأحباب والتلهي بملامسة الأبدان الناعمة الغضة ومغازلة الغنجات منهن ومفاكهة الخنثات ومحادثتهن والتوهم لمحاسنهن فربما ذكر الشاب الشبق لمن يحبه فينعظه التوهم أو يمذيه وذلك عندي مما يحقق بعض أقوال الهند في الوهم، وينفع للجماع أيضا الأشياء الحارة الرطبة مثل الشقاقل المربى والزنجبيل والجرجير والبصل والجزر وبزورها كلها والعنصل المشوي والهليون المسلوق والقلقاس المطحن بالأفاويه والبيض النيمبرشت ولحوم العصافير الذكران والديكة وخصاها وخصى العجاجيل وتدهين الذكر فيما قالت

٢ للمألوفة : المالوفة Ox, Is || ولزومها Ox : – Is || الأشياء Ox : الموساه Is ٣ وتعجب بها النفس Ox :
الأغاني ٦ Is الحماعه : Ox الجماع || Is اذ : Ox إذا ٥ Is فى : Ox إلى || Is – : Ox لها || Is الانفس
Ox : فربما || Is منهس + : Ox الخنثات || Is ملاومه : Ox مغازلة ٧ Is الملاهى والاعانى : Ox والملاهي
الشقاقل || Is من الحماع : Ox للجماع ٩ Is قول : Ox أقوال || Is و : Ox أو || Is مس : Ox لمن ٨ Is وىما
١١ Is الهسلون : Ox الهليون || Is ىررها : Ox بزورها ١٠ Is الحرحم : Ox الجرجير || Is الششماقل : Ox
فيما قالت الهند ١.٥٨–١١ Is حضا الذكران من : Ox خصاها وخصى || Ox التيمرشت : Is النيمبرشت
Ox : – Is

27 This reference must be seen in the full context of Ṭabarī's Ayurvedic sources, as many of the foregoing recommendations are in fact traceable to Indian theories on 'virilification' (*vājīkaraṇa*). Thus, a relevant passage from the *Aṣṭāṅgahṛdayasaṃhitā* reads: "Anointing, massaging and bathing the body; wearing different kinds of beautiful garlands, dresses and jewels; associating with like-minded, obliging friends who are well

What gladdens the heart and arouses (a desire) for sexual intercourse is keeping (the body) clean; bathing with warm water; having an appetite for familiar, habitual foods and drinks; clinging to (the smell of) aromatic plants and perfumes; wearing colourful clothes and (other) things which delight and astound the soul, and which fill it with joy and wonder; drinking in the company of friends; beholding clear and dear faces; minding dental hygiene; applying kohl; rubbing the body with suitable oils; observing animals that copulate; thinking about the different kinds of coitus, and looking at poetry or books in which this is described or depicted; listening to songs and melodies which kindle a yearning for the wives and lovers; enjoying the touch of soft, tender bodies; flirting, bantering and chatting with coquettish, sensual women, and imagining their charms, for sometimes a lustful young man need no more than evoke the mental image of his beloved in order to get an erection or to ejaculate—this is my opinion, which is confirmed by certain teachings of the Indians about imagination.[27] Also conducive to sexual intercourse is (eating) hot and moist stuff, like preserved baby carrots, ginger, rocket, onions, carrots and all their seeds, roasted squills, cooked asparagus, taro milled with spices, poached eggs, the meat of male sparrows, cocks and their testicles, as well as the testicles of calves; (further), according to what the Indians say, anointing the cold, weak, flaccid penis with oil in which §22

versed in music, poetry and storytelling; swimming in a pond amidst lotuses; watching bees humming about, drunk by the nectar of such flowers; spending time in green forests on mountain slopes, under different kinds of trees that gladden the eye, with the call of a cuckoo pleasing the ear, in a seasonal climate that is agreeable to the body and that gives happiness to all those who are present; chewing betel; drinking wines; having an attractive mistress on the lap, in a moonlit night—these and *any other thing which the mind desirously imagines* are aphrodisiacs"; further: "Every object perceived by the senses may yield happiness and love, but what if they all converge in the body of a woman?", see VāgAṣṭ 3/418f. For similar expositions cf. SuSaṃ 2/251 and CaSaṃ 3/92ff.

الهند بالدهن المغلى بالأفربيون إن كان مسترخيا باردا فشلا، وتنفع له الأشياء التي تنفخ أيضا وشرب ألبان اللقاح لا سيما العوذ المطافيل وهى الحديثات الأسنان القريبات النتاج.

§23 فأما ما يورث الملالة والفتور فتعذر الشهوات وقلة المطاعم وخشونتها وأن تكون حارة يابسة أو باردة يابسة جدا وأشد من ذلك الفقر والترحات والأوجال والهم والكد والفِكَر الرديئة ودوام الركوب والركض وتتابع التخم والسكرات فإن ذلك يورث الأسقام والفترة وارتداد القوة وازدياد ضعفها وفلولها وانطفاء الشهوة وموتها.

§24 وقد قال أرسطاطاليس: إن المطعم أو المشرب إذا كثر على المعدة أطفأ نارها فجرت الأغذية في البدن غير نضيجة وصار ذلك نقصانا للبدن لأنها تورث الفترات والداء، وذلك بمنزلة الشجرة التي إذا كثرت عليها المياه عفنتها وإن قلت جففتها أو كالسراج الذي إن كثر دهنه أو قل جدا انطفأ.

§25 ومما يقطع الجماع ويلهي عنه وعن ذكره التشاغل بالأعمال المهمة وقراءة الكتب الملهية عنه والنظر في الوجوه السمجة ومس الأبدان السهكة الخشنة وتذكر الدار الآخرة، كما صنع أعرابي بلغنا أنه

الأسنان ‖ Is منها + :Ox المطافيل ‖ Is و + :Ox اللقاح ٢ Is سمح منه :Ox تنفع له ‖ Is – :Ox باردا ١
الفقر والترحات ... الشهوة ٦–٤ Is كله فيها + :Ox ذلك ٤ Is بان :Ox وأن ٣ Is – :Ox القريبات
السهكة ١١ Is – :Ox ومما يقطع ... الملهية عنه ١٠ Is – :Ox وقد قال ... انطفأ ٩–٧ Is – :Ox وموتها
Is السمكه :Ox الخشنة

---

28 The external application of spurge-enriched oil—or any oil for that matter—to treat erectile dysfunction, cannot be verified in any of Ṭabarī's Ayurvedic sources, where almost all relevant prescriptions are designed to be ingested; in Indian *erotic* literature, on the other hand, such manual procedures, often involving oil, are quite commonly recommended (e.g. VatKam 169), but there is no record of any translations of these texts from Sanskrit into Arabic. It is therefore more likely that the above statement of Ṭabarī traces back to a piece of oral information.

spurge has been boiled;[28] useful moreover are things that bloat, as well as drinking the milk of she-camels, especially *ʿūḏ maṭāfīl*, meaning those that are young in age and that have recently given birth.[29]

As for those (things) that cause languor and lassitude, (they consist in) denying (physical) needs, and eating little or crude food, (especially) if the latter is hot-dry or cold-dry to a high degree; yet worse than this is living in poverty, distress, fear, or sorrow; labouring; having bad thoughts; constantly riding or running; and suffering from repeated (episodes of) indigestion or intoxication—what is then left are illnesses and exhaustion (due to) a slow erosion and progressive retreat of strength: a dying down and (eventual) death of (all) desire. § 23

Aristotle already said: if too much food or drink burdens the stomach, it stifles its fire;[30] then the nutriments flow into the body uncooked, which leads to physical deterioration in the wake of exhaustion and illness. This is just like a tree which rots if it gets too much water, and withers if it gets too little; or like a lamp which goes out if it has too much oil or not enough. § 24

What deters and distracts from having or (even) contemplating sexual intercourse is being occupied by important work; reading books that have nothing to do with it; seeing ugly faces; touching smelly, rough bodies; and recalling the final abode. This is precisely what a Bedouin did, who, so we are told, § 25

29 Ṭabarī's explanation of *ʿūḏ maṭāfīl* is fairly accurate. More specifically, *ʿāʾiḏ* (pl. *ʿūḏ*) is a female mammal "*that has recently brought forth* [...] or *that has brought forth within seven days*, because her young one has recourse to her for protection [*ʿāḏa bihā*] [...] or *that has brought forth within ten days* or *fifteen days* or *thereabout*, after which she is called *mutfil* [!]"; the latter (pl. *maṭāfīl*) is a female "*having a young one* or *youngling with her* [...] or *that has recently brought forth*". In the present context, strictly speaking, *ʿūḏ maṭāfīl* are she-camels (*liqāḥ*) that have recently brought forth and have their young ones with them; for the two terms and their somewhat tautological combination see LaLex 5/1861a and 2193a.

30 This is likely to be a misattribution; cf., however, AriDA 70 f. (*Περὶ ψυχῆς*): "It is necessary that all nourishment (*τροφή*) should be capable of being digested, and digestion is promoted by heat (*θερμόν*)".

تمكّن من امرأة فلما صار بين شُعْبتيْها ونظر إلى رَكَبها تَذكّر الدار الآخرة وقال: والله امرءًا باع جنةً عرضها السموات والأرض بفِتْر فيما بين نخذيْك لَقليل الخبرة بالمِساحة، ونزل عنها.

§26 وكذلك إن تفكر الرجل في الأشياء التي تحدث بين المجامعيْن عند الجماع مما تعافه النفس وتعزب عنه وتتفرد منه، مثل العِبادي الذي تَمكّن من امرأة ففكر في شيء مما ذكرنا ففترت شهوته واسترخى عِجانه فقالت له المرأة: قُم يا خَيّاب، فقال لها: الخياب مَن فتح جِرابَه ولم يَكِّلْ.

في المضرات الحادثة من السهر والشهوة
وفرط الإسهال وإدمان الباه وكثرة إخراج الدم
وهو الباب الذي ألفنا الكتاب لأجله

§27 إن الدم به تكون تربية الجنين في الرحم ومنه اللبن وبه يكون الفرح والنشاط واحمرار الوجه والمفاكهات والغزل وإليه تستحيل الأغذية كلها أولا فإنها إذا نضجت في المعدة حسوا وكيلوسا

---

١ تمكن من امرأة فلما صار بين شعبتيها و Ox: وحد بسن سعب امراه فلما Is ‖ وقال والله Ox: فقال ان Is ٢ نخذيك Ox: رحلسك Is ‖ الخبرة Ox: البصر Is ‖ و Ox: ىم Is ‖ عنها Ox: وتركها Is ٣ وتعزب Ox: او ىعوف Is ٤ وتتفرد منه Ox: – Is ‖ ذكرنا Ox: ذكر Is ٥ لها الخياب Ox: العبادى Is ٦ السهر و Ox: – Is ٧ إدمان الباه وكثرة إخراج الدم Ox: كىره اخراج الدم وادمان الباه Is ٨ وهو الباب الذي ألفنا الكتاب لأجله Ox: – Is ٩ تكون تربية: يكون تربية Ox, ىكون برىسط Is ١٠ المفاكهات والغزل Ox: الفاكهات والهزل Is ‖ فإنها Ox: فان Is ‖ المعدة Ox: + وصارت Is ١٠–٦٢.١ وكيلوسا جرت Ox: ساحدت Is

---

31 This is the garden (*ǧanna*) of Paradise; the whole expression is Qurʾānic, cf. 3:133 … *wa-ǧannatin ʿarḍuhā s-samawātu wal-arḍu*.

32 I do not know Ṭabarī's source for this anecdote. More than 300 years later, the Ḥanbalite preacher and polymath Ibn al-Ǧauzī (d. 597/1201), in his *Ḏamm al-hawā* or "Rebuke of Passion", transmits a variation of the same story on the authority of the traditionalist Ibn al-Marzubān (d. 309/921) < Isḥāq ibn Muḥammad al-Kūfī < al-ʿUtbī, see ǦauḎH 195 no. 715.

had taken possession of woman; when he was between her legs and saw her pubic hair, he recalled the final abode and said: "By God! A man who sells a garden, wide as the earth and the skies,[31] for the span that is between your thighs, such a one indeed has little knowledge of measurements!" And he got off her.[32]

Equally (discouraging) for a man is to think about the things that can hap- §26
pen between two lovers during sex—(things) which repel the soul, (things) which it shuns and from which it withdraws. So it was that al-ʿIbādī,[33] having taken possession of a woman, thought about one of those things we just mentioned, whereupon his desire abated and his perineum relaxed; and when the woman said to him: "Get up, you loser!" he replied: "A loser is somebody who opens his bag without squeezing it!".[34]

On the damages resulting from insomnia, passion, excessive purging, sex addiction, and overmuch bloodletting —it is because of this chapter that we wrote the book[35]—

Blood makes the foetus grow in the womb; milk (comes) from it; joy, energy, §27
blushing, playful talk and love poetry are (associated) with it; and all food is first transformed into it—for when (food) has been cooked in the stom-

---

33 The identity of al-ʿIbādī (lit. "a member of the [Christian] tribe of ʿIbād"), who plays the 'hero' of this otherwise irretrievable story, remains obscure. It is tempting to associate him with Ṭabarī's famous colleague Ḥunain ibn Isḥāq *al-ʿIbādī* (d. 260/873), who also served as court physician to the caliph al-Mutawakkil; yet whilst Ḥunain's relationship with power was not free of trials and tribulations, his (literary) implication in a story of that nature would presuppose, on the part of Ṭabarī, a wilfull (and risky) act of open disrespect if not hostility, for which there is no tangible justification. Another possible, though much more remote candidate is the pre-Islamic poet ʿAdī ibn Zaid *al-ʿIbādī* (d. c. 600 CE), but here again the link is completely hypothetical.

34 "A loser is somebody who opens his bag without squeezing it" (*al-ḫaiyāb man fataḥa ǧirābahū wa-lam yaktul*), that is a man unable to ejaculate semen.

35 It is not entirely clear what textual range Ṭabarī intended to cover here by the term "chapter" (*bāb*), as subsequent headings are not necessarily authorial—considering the topics that follow, and the way in which they are presented, I would guess that he meant to include §§27–68.

جرت إلى الكبد في مجار رقيقة فصارت في الكبد دما ثم انقسم ذلك في أعضاء البدن كلها وصار قوةً وغذاءً لها، فالإكثار من إخراجه والحمل عليه يعطش الأعضاء ويحبس عنها رزقها وأقواتها وينشف ماء الشباب ويغير نضرته ويسكن سورته ومزاجه، وبيان ذلك أنه إن أفرط مفرطٌ في إخراجه غشي عليه وجف ريقه وغشيه السمادير والتهاويل، والحزم في تجنب ذلك والهرب منه وألا يخرج أحدٌ دما إلا إذا خاف معرّته وهيجانه وبعد أن تحاول تسكينه بالرمان الحلو والعدس المقشر والعناب وما أشبهها فيدافع بذلك ما أمكن وقدر يوما فيوما ويؤخر الأوقات التي اعتاد إخراجه فيها حتى يصير ذلك عادة تجري عليها الطبيعة وتسكن إليها.

§28 وكذلك يفعل السهر الطويل فإنه يهزل البدن وينشف الرطوبات ويغير اللون ويورثه الفترات، وذلك موجود في الناس كلهم وفي الدواب القوية فإنه إن حِيلَ بينها وبين النوم عدة ليال ضعفت وإن حيل بين الرجل وبين النوم عدة ليال غارت عيناه وتغير لونه وفترت شهوته واعتل بدنه كله، والاعتدال في السهر والنوم يرطب البدن ويورث النعمة والنضارة ويظهر الدم ويشرق اللون ويشهي الطعام وتستريح به النفس وتهدأ الأعضاء.

§29 فأما الإسهال الكثير فإنه ضار رديء يخدش الأمعاء ويسحجها ويرققها ويذيب القوى ويحدث العشى حتى يترك ذا الجَلَد والأيْد وهو لا طُرُق به ولا نهوض فإن زاد على ذلك حتى يفرط فيه ويلح على البدن لم تؤمَن معرّتُه وشرّتُه، ولذلك قالت الهند: إن النجوَ عماد من أعمدة البدن فإنه إن أصابت رجلا هيضة أو أسهله بطنه إسهالا كثيرا هدّه وصرعه، ولهذا المعنى منع أبقراط الحكيم

١ صار Is : صارت Ox ٣ مزاجه Ox : مزجه Is ٤ غشيه السمادير والتهاويل والحزم في تجنب Ox : عسته السمادىرى الـماوىل فالحزم من ىحت Is || الهرب Ox : بالقرب Is ٥ ألا Ox : ان لا Is ٦ المقشر Ox : – Is || التي Ox : + قد Is ٧ فيها حتى Ox : فسما Is || تجري : يجري Ox, ىحرى Is || الطبيعة Ox : الطبعه Is ٨ اللون Is : البدن Ox ٩ القوية Ox : + ايضا Is ٩–١٠ ضعفت ... ليال Ox : – Is ١٠ عيناه Ox : عسنه Is ١١ البدن Ox : الحسم Is ١٢ تستريح Ox : استريح Is || تهدأ Ox : تهذا Is ١٣ الأمعاء Ox : المعا Is ١٤ ذا Ox : – Is || والأيد Ox : – Is ١٥ شرته Ox : سره Is || الهند Ox : + و Is || أعمدة Ox : عمد Is ١٦ أو أسهله بطنه Ox : واسهلته Is || هده و Ox : – Is || المعنى Ox : فالمعنى Is || أبقراط Ox : ىقراط Is

ach to a broth and chyle, it flows through thin channels to the liver, where it becomes blood which is then distributed to all organs of the body, giving them strength and nourishment.[36] Therefore, overmuch bloodletting—and its advocation—parches the organs, deprives them of their livelihood and provisions, dries up the juice of youth, spoils its bloom, and dulls its force and temper. This is clearly visible in someone who practises bloodletting excessively: he swoons, his saliva dries up, and he is embroiled by visual disturbances and phantoms, all of which must resolutely be avoided and shunned. (In fact), nobody should extract any blood unless he fears its surge and agitation, and (even then only) after having tried to calm it with sweet pomegranates, peeled lentils, jujubes, and the like. The aim is to refrain (from bloodletting) as much as possible and to defer, day by day, the habitual times of doing so, until this (deferral) becomes itself a habit to which nature conforms and on which it settles.

The same is true for prolonged insomnia, which emaciates the body, dries up (its) moistures, pales the complexion, and leaves fatigues. This is found in all humans and strong beasts of burden—when the latter are prevented from sleeping for several nights, they become weak; and when a man is prevented from sleeping for several nights, his eyes sink in, his complexion pales, his vigour diminishes, and his whole body falls ill. The (right) balance between sleeping and waking (however) moistens the body, leaves wellness and freshness, reveals the blood, illuminates the complexion, whets an appetite for food, calms the soul, and soothes the organs. § 28

As regards frequent purging, this is harmful and dangerous: it scratches and scrapes and thins the bowels, melts away powers, and leads to nightblindness, such that a person is (gradually) deserted by stamina and strength, and (left with) no escape nor (means of) resistance; and if he still insists on this (habit) of abusing the body, his crime and malice will not remain unpunished. This is why the Indians say: defecation is one of the basic functions of the body; if a man is struck by vomitive diarrhoea, or excessively purges his belly, it will fell and crush him.[37] By the same token Hippocrates the sage § 29

---

36 On the generation and movement of blood see UllIM 58 and 64 f.

37 On defecation as a natural urge see e.g. CaSaṃ 1/146 f.; on the debilitating effects of excessive purging see ibid. 6/151; on the combination of diarrhoea and vomiting as a potential cause of death and therefore a reason to reject the patient see MāNid 18.

منه وقال: الدواء لا من فوق ولا من أسفل والدواء من فوق ومن أسفل، وذلك أنه إذا لم يكن في البدن سقم وفضول مجتمعة مضرة فالإقدام على شرب الأدوية المسهلة غَرَرٌ وضَرَرٌ لأنه إذا لم يجد الدواء فضلة يخرجها عطف على الأعضاء الباطنة نخدشها وجرد المِعى، والقصد في ذلك بُعيد الشتاء وقُبيل حَمّارة الصيف وحرارته تجفف البدن وتغسل الأوساخ وتخرج الأثفال وتفتح السدد وتوقد النار الغريزية، والإفراط في جميع ما ذكرنا قاتل.

§30 وكذلك الفرح المفرط والحزن المفرط الغالب يقتلان والخوف الشديد أيضا يقتل، أما الفرح المفرط فإنه يقتل لانتشار الحرارة الغريزية عن القلب فيبرد القلب وأما الحزن والخوف فإن الحرارات تهرب منهما إلى القلب حتى تختنق فيه فتطفأ.

§31 والإكثار من الجماع أيضا له غوائل كثيرة ومضرات غير منكرة فإنه ينهك البدن ويذيب المخاخ ويرخي الأعضاء ويفسد الحواس ويغير اللون ويمتص ماء الشباب ويذهب النشاط ويمرض القوة المدبرة للأبدان، ولذلك صار الخصيان ومن لا يجامع أطول أعمارا، وأنّا لنجد ذلك من فعل الجماع ظاهرا في الفوالج وغيرها من عظام الدواب فإن الفالج إذا أكثر الضراب غارت عيناه ورقت عنقه وسكن مرحه وسقط سنامه، والقصد فيه يجفف البدن وينشط النفس ويصفي الحواس ويشهي

١ والدواء من فوق ومن أسفل Ox : – Is || إذا Ox : – Is ٢ فضول Ox : فصوا Is || غرر Ox : عزز Is
٣ وجرد المعى Ox : – Is ٤ بعيد الشتاء وقبيل Ox : بعد السسا وفسل Is || حمارة : حمار Ox, حسرات
سار : Ox النار الغريزية || Is السدود : Ox السدد ٥ Is الرديه + : Ox الأثفال || Is – : Ox وحرارته || Is
Ox قاتل ايضا : Is أيضا يقتل || Ox فانه يقتل لانتشار : Is والحزن المفرط الغالب يقتلان ٦ Is الغرره
فيه وطفا Ox, فيها فيطفا : فيه فتطفأ || Is فقرب منها : Ox تهرب منهما ٨ Is لانتشاب : Ox لانتشار ٧
: Ox للأبدان ١١ Is بالنساط : Ox النشاط || Is اللسن : Ox اللون ١٠ Is – : Ox ومضرات ٩ Is
١٣ Is حفت عنه : Ox رقت عنقه || Is المابح : Ox الفالج || Is الفوابح : Ox الفوالج ١٢ Is بالابدان
Is سامه فالقصد : Ox سنامه والقصد || Is مزجه : Ox مرحه

38 HippLi 9/104 f. = HippHe[3] 80 (*Περὶ τροφῆς*, cf. FiCH no. 61): “Médication évacuante par le haut, par le bas, ni par le haut ni par le bas”.

dismissed this (practice), saying: a drug does not (work) from above or from below but rather from above and from below[38]—meaning that if the body is free from illness and no noxious residues are gathered (in it), the administration of purgative drugs is very risky and harmful, because then the drug, finding no residue to expel, turns against the internal organs, scratches them, and abrades the guts. The right time to do it is between the end of winter and the furnace of summer, (for intense heat already) dries the body, washes out filth, removes dregs, opens obstructions, and ignites the inborn fire. Excess in all mentioned respects is (potentially) fatal.

Excessive joy or excessive, overwhelming sadness may equally be fatal; also extreme fear. As for excessive joy, it may kill by diffusing the innate heat away from the heart, which then cools off; as for sadness and fear, all (innate) heat flees from them into the heart, where it eventually chokes to death.[39] § 30

Overmuch sexual intercourse also implies many hazards and irrefutable damages, for it wears the body out, melts the (spinal) marrows, slackens the organs, corrupts the senses, pales the complexion, sucks up the juice of youth, takes away (all) energy, and sickens the governing faculty[40]—this is why eunuchs and abstainers live longer. We can clearly observe the (negative) effects of copulation in Bactrian camels and other large beasts of burden: if the male is frequently put to stud, its eyes sink in, its neck becomes thin, its briskness fades, and its humps collapse. Practised in moderation, (intercourse) dries the body, enlivens the soul, clears the senses, and whets § 31

39 For arguably the best indigenous discussion of *psychological* influences on the behaviour of the innate heat and, thereby, on physiological and pathological processes see BuḫAD 164–199 (5th/11th century).

40 The faculties (*quwan*) are speculative entities on whose integrity and interaction all physiological processes depend. In the present text, Ṭabarī explicitly registers and classifies seven faculties (see § 68), though their number was probably considered to be higher already in his days; about a century later, in al-Maǧūsī's medical encyclopedia, we find a full-blown exposition of the facultative system (for an excellent English summary see UllIM 60 ff.). Ṭabarī does mention the three basic faculties: natural (*ṭabīʿīya*, seated in the liver), animal (*ḥayawānīya*, seated in the heart) and psychical (*nafsānīya*, seated in the brain); he further mentions the attractive (*ǧāḏiba*), digestive (*hāḍima*), excretory (*dāfiʿa*) and retentive (*māsika*) faculties, all of which are secondary exponents of the natural faculty. As regards the "governing faculty" (*al-qūwa al-mudabbira*), mentioned here in isolation, this is an exponent of the psychical faculty; it represents, on the one hand, the rational part of the soul (τὸ ἡγεμονικόν) and, on the other hand, regulates what we today call the autonomic nervous system.

الطعام، وفي قطعه ومنع الطبيعة شهوتها إذا هاجت فتور البدن والجزالة وفلول حده وعواقب رديئة وبخارات فاسدة مكروهة ترتفع من المني المحتبس إلى الدماغ فتسقمه.

§32 وقد ذكر بقراط الحكيم أن أبدان الناس يبليها ويهرمها خمسة أشياء: كثرة المجامعة وطول السهر والهم والتعب الدائم وإدمان الصوم، وقد رأينا تتابع الأسقام وكدور الأيام مما يبلي ويسوق إلى الفناء.

في تدبير البدن في الربيع

§33 إن الربيع زمان معتدل شبيه بالهواء والدم ممزوج، ينفع فيه كل شيء معتدل القوى وَسَط مثل الفراريج والطيهوج والدراج والبيض النيمبرشت والخس والهندباء والجرجير ولبن المعز والضأن، ولا يكره فيه كثرة الجماع والحركة وإسهال البطن وإخراج الدم ودخول الحمام.

في تدبير البدن في الصيف

§34 فأما الصيف لحار يابس، ينبغي أن يتوقى فيه كل شيء حار من الأطعمة والأشربة والأدوية والأفاويه الحارة والامتلاء لئلا تنطفئ الحرارة ويؤكل كل بارد من الأطعمة والأغذية مثل لحوم العجاجيل مطبوخا بالخل والفراريج المسمنة بدقيق الشعير وتدبر بماء الحصرم والتفاح وحماض الأترج والإجاص والرمان الحامض وأن يؤكل فيه البيض النيمبرشت وتكون رياحينه وبقوله وفواكهه وأدهانه باردة كلها، وأن تقل فيه الحركة والجماع وإخراج الدم ويقل اللبث في الحمام ويستعمل فيه القيء لأن فضول البدن ترق في الصيف وتطفو فوق المعدة ولا تستعمل فيه الغرغرة ولا الإسهال إلا عند الضرورة والحاجة إليه.

---

٢ مكروهة Ox : – Is || المني Ox : الزرع Is || فتسقمه : فيسقمه Ox, فـسـمـه Is ٣ والهم Is : – Ox ٤ التعب الدائم Ox : نعب دام Is || كدور Ox : كرور Is ٦ ممزوج Is : ممزج Ox || شيء Is : – Ox || وسط Ox : واسط Is ٧ النيمبرشت : النيمرشت Ox, الـنـمرسـب Is ٨ إسهال البطن Ox : الاسهال Is ٩ في تدبير البدن Ox : الـتـدبـر Is ١١ تنطفئ Ox : نمـلى Is ١٢ تدبر بماء Ox : نعصـر ما Is ١٣ النيمبرشت Is : النيمرشت Ox ١٣–١٤ رياحينه وبقوله وفواكهه Ox : وربـاحـنـه وفـواكهه نـقوله Is ١٥ فيه[1] Ox : فى Is || فوق Ox : + فم Is ١٦ لا Ox : – Is || والحاجة إليه Ox : – Is

an appetite for food. To refrain from it, and (thereby) to deny nature (the satisfaction of) a stirring desire, enfeebles the body, (blurs) the wit, and blunts the edge—among (other) nasty consequences, (such as) putrid, odious vapours which arise from suppressed semen to (enter) the brain and make it ill.

Hippocrates the sage mentioned that five factors contribute to the dwindling and ageing of the human body: frequent sexual intercourse, prolonged sleeplessness, sorrow, constant labouring, and obstinate fasting.[41] We have (also) noticed that successive illnesses and gloomy days decline (the body) and accelerate (its) destruction. § 32

On how to look after the body in spring

Spring is a temperate, mixed season which is akin to air and blood. In spring, one benefits from every thing that is of moderate, medium strength, such as chicks, partridge, francolin, poached eggs, lettuce, wild chicory, rocket, and the milk of goats or sheep. In this (season), frequent sexual intercourse, exercise, purging the belly, bloodletting, and visiting the bathhouse are not at all undesirable. § 33

On how to look after the body in summer

As for summer, it is hot and dry. In summer, one should be wary of every food, drink, drug or spice that is hot; (avoid) repletion for fear of stifling the (innate) heat; eat any cold foodstuffs, such as veal cooked in vinegar, chicks that have been fed with barley meal and prepared with a juice of unripe grapes, apples, citron sap, plums and sour pomegranates; eat poached eggs; and (turn to) all aromatic plants, greenstuffs, fruits and oils that are cold. In this (season), exercise, sexual intercourse, and bloodletting should be reduced, as well as the length of stay in the bathhouse; vomiting should be practised, because in summer bodily residues are thin and float on top of the stomach; gargling (however) should not be done, nor should (the belly) be purged unless it is absolutely necessary. § 34

---

41 This appears to be an inverted reformulation of the original Hippocratic statement "L'habitude pour les choses qui entretiennent la santé: le régime, le couvert, l'exercice, le sommeil, le coït, le moral", see HippLi 5/352 f. (*Ἐπιδημίαι*, cf. FiCH no. 19).

في تدبير البدن في الخريف

§35 الخريف بارد يابس، فينبغي أن يتوقى فيه كل طعام وشراب بارد يابس ويؤكل فيه كل ما كان حارا لينا رطبا مثل الفراريج والخراف والعنب الحلو والشراب الرقيق ويجتنب كل ما يولد السوداء، وتكون الحركة والجماع والغرغرة فيه أكثر منها في الصيف وأقل منها في الشتاء والربيع ويتعهد فيه الحمام ويتمرخ فيه بالخيري وما اعتدل من الأدهان فإن احتيج إلى القيء كان ذلك وسطَ الشهر أو في آخره لأن الفضول تجتمع في الإنسان في هذيْن الفصليْن ويسهل البطن بأفثيمون وأغاريقون ويقل إخراج الدم فيه.

في تدبير البدن في الشتاء

§36 قال جالينوس: إن فضول البدن تقل وتخفى في الشتاء لأن البرد يجمدها، وتكثر في الصيف لأن الحر يذيبها ويظهرها.

§37 وقال أبقراط الحكيم المقدم: إن البطون في الخريف والشتاء حارة والنوم فيهما كثير لطول الليل فينبغي أن يكون الطعام فيهما أكثر، والبطون في الصيف والربيع باردة فينبغي أن يؤكل فيهما دون ما يؤكل في الخريف والشتاء.

§38 وأن يؤكل في الشتاء الأشياء الحارة مثل فراخ الحمام والعصافير وحولي الضأن والبقول والتوابل الحارة والكباب والتين والكراث والجوز، ولا يضر فيه كثرة الجماع والحركة وتستعمل الجوارشات

١ في تدبير البدن Ox : الـدبـر Is ٢ فينبغي Ox : يـنـغى Is ٣ رطبا Ox : – Is ‖ الخراف Ox : الحرفان Is ‖ يجتنب كل Ox : يـحـتـب Is ٤ منها¹ Ox : منه Is ‖ منها² Ox : مما Is ٥ احتيج Ox : احتاج Is ‖ ذلك Ox : + فى Is ٦ أغاريقون Ox : غاريـقون Is ٧ فيه Ox : + ان سا الله Is ٨ في تدبير البدن Ox : الـدبـر Is ٩ وتخفى Ox : – Is ‖ يجمدها وتكثر : يخمدها ويكثر Ox, يحمدها ويكثر Is ١١ أبقراط Ox : بقراط Is ‖ فيهما : فيها Ox, فيها Is ‖ الليل Ox : النهار Is ١٢ الصيف والربيع Ox : الربيع والصيف Is ١٣ فيهما² Ox : فيها Is ‖ يؤكل Ox : نوكل Is ١٥ والكراث : + والتين Ox, – Is ‖ الجماع والحركة Ox : الحركه والجماع Is ‖ الجوارشات Ox : الجوارسنات Is

On how to look after the body in autumn

Autumn is cold and dry. Accordingly, one should be wary in autumn of every food or drink that is cold or dry; (instead), one should consume all hot, soft and moist stuff, such as chicks, lambs, sweet grapes, and thin wine, avoiding anything that generates black bile. In this (season), exercise, sexual intercourse, and gargling should be practised more (frequently) than in summer and less (frequently) than in winter or spring; the bathhouse should be (visited) regularly, the skin rubbed with gillyflower or another temperate oil; if vomiting becomes necessary, one should (provoke) it around the middle or towards the end of the month, because it is during these two periods that residues gather in the human (body); the belly should be purged with dodder and agaric, but bloodletting should be reduced. § 35

On how to look after the body in winter

Galen said: in winter, bodily residues are few and hidden, because the cold freezes them; in summer, they are many, because the heat melts and exposes them.[42] § 36

Hippocrates the ancient sage said: in autumn and in winter the bellies are hot, and sleep is ample because the nights are long—accordingly, one should have more food then; in summer and in spring the bellies are cold—accordingly, one should eat less during these two (seasons) than in autumn and in winter.[43] § 37

In winter, one should eat hot things, such as young pigeons and sparrows, sheep yearlings, greenstuffs, hot condiments, spit-roast meat, figs, leek, and walnuts. In this (season), frequent sexual intercourse, exercise, as well as the § 38

42 The closest, albeit fairly incongruous, Galenic statement here is the following: "Quemadmodum autem aestate et humores discuti et vires dissolvi contingit, sic et hieme contraria eveniunt, tum humores intus manere tanquam in latibulis delitescentes et vires servari ac permanere robustas", see GalKü 17B/514 (*Περὶ τῶν Ἱπποκράτους ἀφορισμῶν*, cf. FiCG no. 101).

43 HippLi 4/466–469 (*Ἀφορισμοί*, cf. FiCH no. 13): "En hiver, et au printemps [!], le ventre est naturellement le plus chaud, et le sommeil le plus long; c'est donc dans ces saisons qu'il faut donner plus de nourriture" and "Pendant l'été et l'automne [!], la nourriture est supportée le plus difficilement, le plus facilement pendant l'hiver, en second lieu pendant le printemps".

والحقن الحارة ويتوقى الإسهال إلا أن يُضطرّ إليه فيغيَّر الهواء ويسخَّنه وتستعمل الأفاويه والأدهان والأشربة الحارة.

في تدبير البدن في الحمام

§39 إن في الاستحمام منافع لأهل كل سن وفي كل زمان إذا كان الحمام معتدلا في حره وكان ماؤه عذبا جاريا ولم يخالف التدبير فيه، وهو أن من كان شيخا أو غلبت عليه البرودة والرطوبة لبث فيه طويلا حتى يتصبب عرقه ومن كان شابا والغالب عليه الحر واليبس لبث فيه أقل بقدر ما يبتل بدنه ويأخذ من رطوبة الحمام ولا يأخذ الحمام من رطوبته، وليعلم أن الحمام مبني على فصول السنة التي لا تؤذي الأبدان فمن حر الربيع الفاتر إلى حر الصيف الفائت ومن برد الخريف الخفيف إلى برد الشتاء الفارس.

§40 ومن صواب التدبير في الحمام أن يلبث الرجل في البيت الأول قليلا ثم يصير منه إلى البيت الثاني فيلبث فيه قليلا ثم يصير منه إلى البيت الثالث وكذلك يفعل فيه إذا أراد الخروج يلبث في كل بيت هنيهة لئلا يهجم من برد شديد على حر مفرط على برد يابس، فإن كان محرورا أو يابس المزاج أسرع صب الماء عليه وخرج، وإن كان شيخا أو بارد المزاج أو كثير الرطوبات أطال اللبث لتتحلل فضوله وترشح رطوباته ويجف بدنه، ويستحب لصاحب البلغم أن لا يستحم إلا على الريق وأن يستنقع في ماء قد طبخ فيه المرزنجوش والشيح والنمام والغار والقيصوم أو يضع يديه ورجليه

---

١ فيغير : بتغير Ox, فـعـىـر Is ‖ يسخنه : ىسخن Ox, ىسحں Is ٢ الحارة Ox : + ان سا الله Is ٣ في تدبير البدن Ox : الـىـدىـر Is ٤ كل[1] Ox : + و زماں Is ٦ عرقه يتصبب Ox : ىـصـب عرڡا Is ‖ الحر Ox : والحرو Is ‖ أقل Ox : فلـىلا Is ٧–٩ الشتاء الفارس ... ولا يأخذ الحمام Ox : – Is ٧ فصول Ox : فضول ١٠ في الحمام Ox : فـىـه Is ‖ منه Ox : – Is ١٠–١١ البيت ... الثاني Ox : – Is ١١–١٢ يابس[1] ... وكذلك Ox : – Is ١٢ لئلا : هنيهة هنيهنة Ox ‖ المزاج يابس Ox : ىـاىـسا Is ١٣ عليه Ox : + وافراغه Is ١٣–١٤ لتتحلل اللبث : لينحلل الليث Ox, لـىـحلل اللتـب Is ١٥ أو Is : و Ox

---

44 The phrase *fa-yuġaiyar al-hawāʾ wa-yusaḫḫanuhū* has been emended and interpreted on the basis of the parallel passage ṬabFir 109,7 f. which reads *fa-yusaḫḫan hawāʾ al-bait* [!] *ṯumma yushal ishālan ḫafīfan*.

use of stomachics and warm enemas do no harm; one should (however) be wary of purging (the belly), unless it is necessary and (only after) having changed the (room) temperature by heating it up;[44] (further recommended is) the use of hot spices, oils and beverages.

On how to look after the body in the bathhouse

Bathing has (many) benefits for people of all ages and in all seasons, provided that the bathhouse is reasonably hot, that its water is fresh and running, and that one does not deviate from the (following) procedure: an old man, or somebody who is dominated by coldness and moisture, stays in it for a long time, until he drips with sweat; a young man, or somebody who is dominated by heat and dryness, stays in it for a shorter time, until his body becomes wet and he has taken moisture off the bath, rather than the bath taking moisture off him. One must know that the bathhouse is constructed in analogy to the seasons of the year, which hold no harm for the body—(ranging) from the balmy warmth of spring to the blazing heat of summer, and from the gentle coolness of autumn to the bitter cold of winter. § 39

The proper conduct of a man in the bathhouse is to stay in the first chamber for a little while, then go from there into the second chamber, where he (also) stays for a little while, and finally go into the third chamber; when he wishes to leave, he proceeds in the same way, staying shortly in each chamber, without rushing from severe cold into intense heat through dry coldness.[45] If he is hot- or dry-tempered, he quickly pours the water over himself and leaves. If he is an old man or cold-tempered or full of moisture, he stays longer, such that his residues dissolve, his fluids ooze out, and his body dries up. A phlegmatic person should visit the bathhouse preferably on an empty stomach, and sit in (a tub filled with) water in which marjoram, wormwood, wild thyme, bay laurel and southernwood have been cooked, or put his hands and feet under (running) water;[46] he should (also) rub himself § 40

45 For virtually all features and aspects of the medieval Islamic bathhouse (*ḥammām*), with the exception of strictly medical implications, still see GroBad *passim*.

46 The two clauses "and sit in (a tub filled with) water … or put his hands and feet under (running) water" have been understood in accordance with the parallel passage ṬabFir 341,1 where the crucial phrase *fa-in lam yakun fī l-ḥammām ābzan* is inserted.

في ذلك الماء ويتمرخ بأدهان حارة ويبخر الحمام بالعود والقسط والقرنفل، وإن كان محرورا لا يدخله إلا بعد هضم الطعام ولا يدخل البيت الحار إن خاف ضعفا ثم يستنقع في ماء قد طبخ فيه البنفسج والورد والنيلوفر أو يضع يديه ورجليه فيه ويأكل حين يخرج وقبل أن تثور به الحرارة ويبخر له الحمام بعود صرف أو بالكافور والزعفران، فأما من كان باردا أو رطبا فينبغي أن يبخر له الحمام بعود مطرًّى أو بالقسط والقرنفل.

في علامات وعلاجات أطباء بابل وغيرهم

§41 قالوا: إن البدن أربعة أجزاء، الأول منها الرأس وما يليه فإذا اجتمعت فيه فضول كان آية ذلك ظلمة العين وثقل الحاجبين وضربان الصدغين ودوي الأذنين وانسداد المنخرين، فمن أحس بذلك فيأخذ الأفسنتين فليطبخه بشراب حلو مع أصول السعتر حتى يذهب نصفه ويتغرغر به كل غداة أو يأكل الخردل بالعسل فإنه إن أغفل ذلك أورثه وجع العين والخنازير والذبحة وأوجاع الرأس.

§42 والجزء الثاني الصدر وما يليه فإذا اجتمعت فيه فضول كان آية ذلك ثقل اللسان وملوحة الفم أو مرارته وحموضة الطعام على رأس المعدة ووجع العضدين والسعال، فينبغي لمن أحس بذلك أن يخفف من طعامه ثم يتقيأ فإنه إن أغفل ذلك أورثه ذات الجنب ووجع الكلية والحمى.

§43 والجزء الثالث البطن وما يليه فإذا اجتمعت فيه فضول كان آية ذلك التقطير ووجع الركبة والقشعريرة والمليلة، فينبغي لمن أحس بذلك أن يسهل البطن ويفرغ الفضول فإنه إن أغفل ذلك أورثه استطلاق البطن ووجع الورك والظهر والمفاصل والبواسير.

---

١ و[1] Ox : او Is ‖ ويبخر ... القرنفل Ox : – Is ٢ ثم Ox : و Is ٣ أو Is : و Ox ٤ فينبغي Ox : ىمعه Is
٥ الحمام Ox : + و Is ٩ فيأخذ الأفسنتين فليطبخه Ox : فلىاخذ الافسىىں فىطىخه Is ‖ مع Is : و
Ox ‖ السعتر Ox : الصعىر Is ١٠ أو Ox : و Is ‖ بالعسل Ox : والعسل Is ‖ الخنازير Ox : الحدالدىں Is
١١ فإذا اجتمعت وما يليه Ox : فاذ احتمع Is ١٢ الطعام Ox : – Is ‖ و[2] Is : او Ox ‖ لمن أحس بذلك
Ox : – Is ١٣ و ذات الجنب Is : – Ox ١٤ اجتمعت Ox : احتمع Is ١٥ القشعريرة : الاقشعريره
Ox, الاقسعرىره Is ‖ فينبغي Ox : + ان Is ‖ فإنه إن Ox : فان Is ١٦ استطلاق أورثه Ox : اووربىه
اصطلاق Is

with warm oils, and fumigate the bath (chamber) with lignaloes, costmary and clove. A hot-tempered (person) should go to the bathhouse only after (his) food is digested, and he should not enter the hot chamber if he fears a weakness; then he sits in (a tub filled with) water in which sweet violets, roses and nenuphars have been cooked, or puts his hands and feet under (running water);[47] he should leave (the hot chamber) before the heat overwhelms him, and at that point have something to eat; and he should have fumigated (the hot chamber of) the bath with pure lignaloes or with camphor and saffron. As for somebody who is cold or moist (by nature), he ought to fumigate the bath (chamber) with perfumed lignaloes or with costmary and clove.

On symptoms and treatments (according to) Babylonian and other physicians[48]

They say: the body (consists of) four parts. The first of these is the head and what lies next to it. Signs of an accumulation of residues in the head are a darkening of the eyes, a heaviness of the brows, a throbbing in the temples, a humming in the ears, and a clogging of the nostrils. He who experiences this (should) take absinthe and savory roots, cook them in sweet wine down to one half, and gargle with it every morning; or he eats mustard in honey. If he neglects that, he will be left with eye pain, scrofula, a sore throat, and headaches. § 41

The second part is the chest and what lies next to it. Signs of an accumulation of residues in the chest are a heaviness of the tongue, a salty or bitter taste in the mouth, acid reflux, pain in the upper arms, and a cough. He who experiences this should eat light food, then vomit. If he neglects that, he will be left with pleurisy, kidney pain, and fever. § 42

The third part is the belly and what lies next to it. Signs of an accumulation of residues in the belly are dribbling (urine), knee pain, shivers, and feverishness. He who experiences this should purge the belly and empty out the residues. If he neglects that, he will be left with diarrhoea, pain in the hip, back and joints, and haemorrhoids. § 43

---

47 Cf. note 46 above with ṬabFir 341,5 (where to read *fīhī* instead of *ṯumma*).

48 §§ 41–44 of this chapter seem to reflect certain teachings of 'Babylonian' physicians, for which concept see p. 26 above; "other physicians" refers to Byzantine, Indian and Persian practitioners, as they are explicitly mentioned in § 45 of this chapter.

§44 والجزء الرابع المثانة وما يليها فإذا اجتمعت فيها فضول كان آية ذلك فتور الشهوة وتبثر الإليتين والعانة، فينبغي لمن أحس بذلك أن يأخذ من الكرفس والرازيانج ومن أصولهما فينقعه في شراب أبيض طيب الريح ثم يأخذ منه في كل غداة ممزوجا بالماء والعسل على الريق ويحتمي من كثرة الأكل فإن من أغفل ذلك أورثه وجع المثانة والكبد وحصر البول والربو.

§45 وذكرت الأعاجم أن ملكا من ملوكها جمع أطباء الروم والهند والفرس وأمرهم أن يصف كل واحد منهم شيئا إذا لزمه واستعمله في أيام السنة نفعه ولم يضره فكان ما اختاره وأشار به الطبيب الرومي الماء الحار وما أشار به الطبيب الهندي الهليلج الأسود المربى وما أشار به الطبيب الفارسي الحرف وقد ينفع كل واحد منها مفردا من أدواء كثيرة، وقالت الفرس: إن من أمسى وليس في بطنه ثقل لم يخف الفالج ووجع المفاصل، ومن أكل في كل شهر سبعة أيام في كل غداة سبعة مثاقيل من زبيب أحمر لم يخف أدواء البلغم وجاد حفظه وكذلك إن أكل بدل الزبيب سبعة أيام أعواد زنجبيل مربى بالعسل أو عودا لطيفا من الوج المربى بالعسل، ومن أكل في كل غداة كل يوم من أيام الشتاء ثلاث لقم من الشهد لم يصبه البرسام شَتْوَتَه تلك ومن أكل كل غداة من أيام الصيف خيارة واحدة لم يصبه البرسام صَيْفَتَه تلك، ومن أدمن شم المرزجوش واستعمل دهنه لم يصبه صداع ولم ينزل في عينه ماء، ومن استعمل أكل الحلتيت أمن حمى الربع، ومن أكثر المشي

١-٢ تبثر الإليتين والعانة :Ox سو على السس والعايه Is ٢ الكرفس :Ox الكورفس Is || أصولهما Ox
الطبيب² || Is – :Ox المربى ٧ Is وان :Ox فإن ٤ Is عذاه :Ox غداة || Is – :Ox منه ٣ Is اصولها :
Is أمسى وليس في || Is مسهما :Ox منها مفردا || Is فمد سمع ,Ox وقد ينقع :وقد ينفع ٨ Is – :Ox
أحمر ١٠ Is – :Ox سبعة أيام في كل غداة | Ox غداة :غداة || Is – :Ox كل¹ ٩ Ox امـ••ـ• و•ـ•ـ•ـ• :
أيام² ١٢ Is – :Ox كل يوم ١٢-١١ Is بعسل :Ox بالعسل² || Is – :Ox مربى بالعسل ١١ Ox – :Is
Is الما :Ox ماء ١٤ Is المرزبحوس :Ox المرزجوش || Is من + :Ox البرسام ١٣ Ox – :Is

---

49 “Foreigners” (*aʿāǧim*) are generally non-Arabs or barbarians—that is people who speak incorrect or incomprehensible Arabic (cf. LaLex 5/1967b–c)—and more specifically Persians; in the present context, the term most probably refers to *Sasanians* and, by extension, to their declared or perceived descendants.

50 It is tempting to associate this presumably legendary tale with the Sasanian king Khosrow I Anushirvan (reg. 531–579 CE) and his court at Ctesiphon.

The fourth part is the bladder and what lies next to it. Signs of an accumu- §44
lation of residues in the bladder are a flagging of desire, and a pustulation of the buttocks and pubic region. He who experiences this should take celery and fennel, including their roots, and soak them in fragrant white wine, which he then drinks, every morning, mixed with water and honey, on an empty stomach; and he (should) refrain from copious eating. If he neglects that, he will be left with pain in the bladder and liver, urinary retention, and asthma.

The foreigners[49] report that one of their kings summoned (three) physicians §45
from Byzantium, India and Persia, and ordered each of them to prescribe something that, if he were to stick to it and use it throughout the days of the year, would serve him well and do no harm; the Byzantine physician chose and advocated hot water, the Indian physician advocated preserved black myrobalans, and the Persian physician advocated gardencress[50]—and (it is true that) each of these (substances), on their own, may be helpful against a variety of illnesses. The Persians say: he who enters into evening without a (sensation of) heaviness in his belly, ought not fear hemiplegia nor painful joints; he who eats, on seven days every month, seven *miṯqāl* of red raisins in the morning, ought not fear phlegm-related diseases, and his memory will improve—the same (applies) if he eats, on seven days, instead of raisins, (some) ginger sticks preserved in honey, or a tender stem of sweet flag preserved in honey; he who eats, in the morning of each winter day, three morsels of honeycomb, is not hit by pleurisy in that winter, (just as) he who eats, in the morning of each summer day, one cucumber, is not hit by pleurisy[51] in that summer; he who keeps smelling marjoram and using its oil, is not hit by headache nor befallen by cataract;[52] he who regularly eats asafoetida resin, is safe from quartan fever; he who often walks around barefooted, is

51 Due to an old and partly graphical confusion between the Persian words *sar-sām* (سرسام) "head inflammation" and *bar-sām* (برسام) "breast inflammation", these terms are often used ambiguously, see the discussion UllIM 28ff. In the above passage, Ṭabarī employs the term *barsām* twice to denote "pleurisy", whereas in §64 the same term appears in a context that rather points to a cerebral condition (scil. "phrenitis"); note, however, that in the latter case the German translator of the text opted for the interpretation "Brustfellentzündung", see ṬabHT 64.

52 For *mā'* (lit. "water") as a technical term (i.q. *ὑπόχυμα*) to denote cataract see MeyTT ١٤٠ (Arabic) = 68 (English) and *passim*.

حافيا أمن النقرس، ومن أخذ كل غداة جوزتين وثلاث تينات مع ورقات سذاب لم يخف يومَه السم بإذن الله تعالى.

في علة الاغتذاء

§46 قال أرسطاطاليس الفيلسوف: إن الإنسان إنما يكون حيوانا بالحس ويكون ناميا ناشئا بالحركة والغذاء، وإن علة الاغتذاء هى الشوق إلى البقاء.

§47 وقال جالينوس: إنه ما دامت في الإنسان حرارة معتدلة ورطوبة غير مفرطة تغتذي بها تلك الحرارة الغريزية فإنه تُرْجَى له الصحة والبقاء ويكون موته بانطفاء الحرارة الغريزية وبالبرودة، فإنه إنما يهرم الإنسان ويبلى بدنه لخصلتين إحداهما هرم طبيعي باضطرار ولا محيص عنه ولا بد منه وكذلك يبس البدن والآخر هو هرم عرضي مثل ما يعرض من الآفات والأمراض.

١ تينات Is : تينا Ox   ١-٢ يومه السم بإذن الله تعالى Ox : السم بومه Is   ٥ هى Ox : هو Is   ٦ غير مفرطة Ox : – Is ‖ بها Ox : به Is   ٧ الغريزية¹ Ox : – Is ‖ فإنه ترجى ... الغريزية Is : – Ox ‖ بالبرودة Ox : بالبرد Is   ٨ يبلى Ox : يهزل Is ‖ إحداهما Ox : احدهما Is ‖ باضطرار ولا محيص Ox : بالاضطرار لا لمحيص Is   ٩ هو هرم عرضي Ox : عرص Is ‖ والأمراض Ox : – Is

53 The latter section of this paragraph, listing a number of semi-sympathetic measures specifically according to what "the Persians say", cannot be substantiated in the sources, as there are virtually no medico-pharmaceutical texts of Pahlavi derivation available for comparison, cf. RhaCB 48–59; considering the nature of these materials, it seems in any case more fitting to interpret them as examples of (eastern) Iranian folkmedical traditions, upon which Ṭabarī is known to have frequently relied, see SchṬab 33 and UllMed 121.

54 This extremely terse 'quotation' is an abridgement of the following parallel passage in the *Paradise of Wisdom* (ṬabFir 114,21–115,3): *inna l-insān innamā ṣāra ḥayawānan bil-ḥiss wa-mufakkiran bil-ʿaql wa-nāmiyan bil-ḥaraka wal-ġiḏāʾ wa-innahū lammā lam yakun lil-insān an yabqā bi-šaḫṣihī ḥtāǧa wa-štāqa ilā an yabqā bi-ṣūratihī fa-ḥtāǧat aṭ-ṭabīʿa li-ḏālika ilā n-nasl fa-lam yaṣil ilā n-nasl illā bin-numūw wa-lā ilā n-numūw illā bil-ġiḏāʾ fa-ʿillat al-ġiḏāʾ hiya š-šauq ilā l-baqāʾ* "Man is a living being endowed with sensation; he is a thinking being through the intellect; and he is a growing being because he moves and feeds. As he cannot last in his individual form, he deeply desires to last

safe from gout; and he who takes, every morning, two walnuts and three figs together with rue leaves, ought not be afraid of poison on that day—with the permission of God the Sublime.[53]

On the reason for self-nutrition

The philosopher Aristotle said: man is a living being endowed with sensation; he grows and develops through movement and nourishment. The reason for self-nutrition, then, is an urge to subsist.[54] §46

And Galen said: as long as a person possesses equable warmth and sufficient moisture, the innate heat nourishes itself from them, and health and survival can be expected for him; his death is caused by an extinction of the innate heat and by coldness.[55] A person ages and his body wanes in two ways: the one is a natural, necessary decline, from which there is absolutely no escape—(progressive) desiccation of the body is similar (in this respect); the other is an accidental decline, such as it happens through plagues and diseases.[56] §47

---

as an image of himself, and to this end nature needs procreation; yet there is no procreation without growth and no growth without nourishment, which is why the reason for (self-)nutrition is an urge to subsist". This passage, in turn, merely serves to introduce a radically curtailed exposition of certain Aristotelian ideas about the 'nutritive function' (*al-qūwa al-ġāḏiya*) of the soul (*ἡ δύναμις τῆς ψυχῆς θρεπτική*), and Ṭabarī soon shifts the emphasis from metaphysics to physiology, as can be expected. The Greek source text for Ṭabarī's paraphrases is book 2 of *Περὶ ψυχῆς* (AriDA 48–107, esp. 54f., 58f. and 62–65).

55 GalKü 1/522f. = GalHe 9 (*Περὶ κράσεων*, cf. FiCG no. 9): "Sed et mortem ajunt animantium corpora ad frigidum siccumque perducere; quippe mortuos *ἀλίβαντας* vocari, quasi nihil humecti in se habentes, utpote tum caloris abitione eo exhaustos, tum frigore rigentes. Quod si, inquiunt, mors talis est naturae, certe vita, quum sit illi contraria, calida est et humida; at vero si vita calidum quiddam atque humidum est, omnino quod illi simillimum temperamentum est, id optimum necessario est"; further GalKü 1/582 = GalHe 47 (op.cit.): "Itaque si mors naturalis caloris est extinctio […]". Galen's views on the function of corporal heat are, however, somewhat ambivalent.

56 GalKü 15/294 (*Εἰς τὸ Ἱπποκράτους περὶ τροφῆς ὑπομνήματα*, cf. FiCG no. 92): "Duobus modis corruptela dicitur, tum ea quae fit, tum ea quae facta est"; further GalKü 7/407 (*Περὶ τῶν ἐν ταῖς νόσοις καιρῶν*, cf. FiCG no. 49): "Morbi corporum accidentia".

في أقدار الأغذية ومنافعها

إن من الأغذية ما هو لطيف ومنها غليظ ومنها وسط، فاللطيف منها يولد دما صحيحا صافيا جيدا مثل الحنطة والفراريج والبيض غير أنها تضعف من استعملها، فأما ما كان من الأغذية غليظا فإنه ينفع المحرورين ومن كثر تعبه قبل الطعام ونومه بعد الطعام ولمن يحتاج إلى أن يكون له جَلَد وبَطْش غير أنها تولد سددا وفضولا غليظة إلا أن يكون الذي يستعملها كثير الحركة قبل الأكل كثير النوم بعده، فأما الوسط من الأطعمة فإنها لا تولد السدد والفضول مثل الأطعمة الغليظة ولا تضعف البدن مثل ما يضعفه الأطعمة اللطيفة وهى مثل الجداء الحولية والمعز والدجاج وما أشبهها. §48

وإنما يصير الطعام خفيفا على المعدة لخصال إما لخفته وإما لاعتداله في حره وبرده أو للطافته أو لأن الرجل يشتهي شهوة شديدة فإما من تعب وقلة غذاء وإما لقوة حرارة المعدة وحركة الجوع، وعلة ثقل الطعام على المعدة إما لثقل الطعام وإما لغلظه أو لبرده أو لصلابته أو للزوجته أو لدسمه أو لأنه يكون كريه الطعم أو مطحنا معمولا من الفطير مثل بعض الحلوى أو لأن المعدة لا تعتاده أو لامتلاء في المعدة والأعضاء من الحدود المحيطة بقوى أشياء كثيرة من الحيوانات والنبت والثمر وغيرها إن كان حيوانا أو ثمرة أو نبتا عظيما غليظا في مضغه. §49

وما كان من الحيوان صلبا يرعى في المراعي اليابسة الصلبة ويكد في الأعمال الصعبة فإنها أصلب وأبطأ استمراءً ويستحيل أكثر ذلك إلى السوداء مثل البقر والإبل العوامل وكذلك §50

---

٢ صحيحا Ox : – Is   ٣–٤ فأما ما كان من الأغذية غليظا فإنه Ox : واما العلـطة فانها Is   ٤ لمن يحتاج Ox : لم يحـتـح Is   ٥ قبل Ox : فلـيل Is   ٧ ما يضعفه الأطعمة Ox : – Is ‖ هى Ox : هو Is ‖ الحولية والمعز والدجاج Ox : والدجاج الحولى والمعز Is   ٩–١٠ أو لأن الرجل يشتهي شهوة شديدة Ox : ان الرحل قد يشتهى الشهوه السديده Is   ١٠ فإما من تعب وقلة غذاء Ox : – Is ‖ لقوة Ox : بقوه Is   ١١ و¹ Ox : – Is ‖ لثقل الطعام وإما Ox : – Is ‖ أو¹ Ox : واما Is ‖ للزوجته Is : للزوجيه Ox   ١٢ يكون Is : – Ox ‖ الطعم Is : المطعم Ox ‖ معمولا Ox : – Is ‖ أو³ Ox : و Is   ١٣ المعدة والأعضاء Ox : معدته واعصايه Is   ١٤ ثمرة أو نبتا Ox : بست او بمر كان Is   ١٥ صلبا يرعى Ox : كلما براعى Is ‖ يكد Ox : بكند Is ‖ الصعبة Ox : الرطبه Is   ١٦ الإبل Ox : + و Is

On the values and benefits of food (items)

Some nourishments are tender, some are tough, and some are in-between. The tender ones generate good, pure, healthy blood—such as wheat, chicks and eggs—, except that they weaken him who uses them. As for tough nourishments, they benefit those who are hot-tempered, those who do a lot of labouring before (their) meal and go to sleep afterwards, and those who require stamina and willpower, except that these (nourishments) produce obstructions and coarse residues if he who uses them does not get plenty of exercise before he eats and plenty of sleep thereafter. As for foods that are in-between, they neither produce obstructions and residues, like the tough ones, nor do they weaken the body, like the tender ones—such (foods) are kid yearlings, goats (in general), chicken, and the like. § 48

(An item of) food may be managed easily by the stomach for (different) reasons: because of its lightness; because of its temperance between heat and coldness; because of its tenderness; or because the man has a fierce appetite, due to either labouring and scanty provisions, or due to intense gastric heat which stirs up hunger. (An item of) food may burden the stomach for (the following) reasons: because of its heaviness, toughness, coldness, hardness, stickiness, or greasiness; because of its horrible taste; because it is a flour-based (product) made from unleavened dough, like some pastry; because the stomach is not accustomed to it; or because the stomach and the surrounding organs are (already) filled and occupied with other stuff—be it meats, vegetables or fruits, (especially) if these were tough and hard to chew. § 49

A sturdy animal that grazes on dry, harsh pastureland and performs strenuous, exhausting tasks, is very hard and slow to digest, and most of it converts into black bile—such is the case with cows, camels and yoke-oxen, § 50

الأيائل والأوعال والغزلان والقطا والقبج وما أشبه ذلك من اللحمان، واللحمان في الجملة حارة رطبة تختلف بالصنعة فيما شُوِيَ منها فإنه يستفيد من قوة النار وما طبخ بالخل والثوم فإنه يستفيد من قوة الخل والثوم.

§51 وكذلك القول في النبت، فإن ثمار شجر السقى أخف من ثمار شجر الجبال والمواضع الحجرية اليابسة الصلبة، ورياحين الجبال والأعذاء ونبتها وعقاقيرها وبزورها وطعومها أقوى من العمرية الأسقاء لأن المياه تكثر عليها فتضعف لذلك قوتها، ولذلك صار سم الحيات التي في الإجام والأنهار سليم والحيات الجبلية والرملية قاتلة.

§52 وكذلك القول في السمك، فإن ما ضعف جسمه ورق جلده وخف مضغته وكان في مياه عذبة جارية فهو أخف مما كان في البحار والإجام أو كان عظيما غليظا، وهى في الجملة باردة رطبة وأنفعها للمرضى ما كان في المياه التي تجري على الصخر والرمل والطري المشوي منها يزيد في الباه ولا سيما إذا كان سمينا.

§53 وكل دابة صغيرة جدا أو نبت أو ثمرة فإنه أول ما يطلع يميل إلى البرودة واللزوجة ويولد غذاء رديئا لأنه بارد لم تستحكم حرارته ولا رطوبته وما كان منها هرما كبيرا جاسئا فإنه رديء الغذاء لغلبة الماء عليه وما كان معتدلا في سنه وقدره وزمانه فهو أخف وألطف وأجود غذاءً، وأحمد اللحمان الجداء والخرفان وأغلظها الجواميس والبقر والإبل والأيائل وسائر البريات والجبليات وأحمد ما يؤكل من بطونها الكبد لأنها تولد دما فاضلا فأما الطحال فإنه يولد السوداء والكرش عصبي

---

١ الأيائل : Is الايايل Ox ‖ واللحمان اللحمان : Ox العزلان Is ٢ منها : Ox منه Is ‖ يستفيد : Ox بسمدونه Is ‖ من قوة : من قوة Ox, – Is ‖ بالخل والثوم : Ox بالثوم والخل Is ٣ من : Ox عان Is ٤ القول : Ox القوا Is ‖ فإن ثمار : Ox وان بمارها Is ٤–٥ الحجرية اليابسة الصلبة : Ox اليابسه الحجريه Is ٥ العمرية : Ox الروصه Is ٦ فتضعف : Ox وبصعف Is ‖ قوتها : Ox قواها Is ‖ سليم : سليمه Ox, Is ٨ ضعف : Ox صغر Is ٩ أخف : Ox احب Is ‖ أو : Ox اذ Is ‖ عظيما : Ox – Is ١٠ للمرضى : Ox للمريض Is ‖ الصخر : Is الصحفة Ox ١١ كان : Ox سهما Is ١٢ يميل : Ox مابل Is ١٣ رطوبته : Ox – Is ١٣–١٤ لغلبة الماء عليه : Ox – Is ‖ جاسئا فإنه : Ox خاسا فانه ايضا Is ‖ منها : Ox – Is ‖ بعد + : Ox Is ١٦ لأنها : Ox لانه Is ١٦–٨٢.١ عصبي بطيء : Ox عسر Is

but also with stags, mountain goats, young gazelles, sandgrouses, partridges, and similar types of meat-giving (animals). On the whole, flesh-meat is hot and moist, and prepared in different ways: when roasted, it profits from the power of the fire; when cooked with vinegar and garlic, it profits from their power.

The same goes for plants. The fruits of irrigated trees are lighter (to digest) than the fruits of trees (that grow) in the mountains or in dry, harsh, stony places; and the aromatic herbs (that grow) in the mountains or in wasteland—(including) their sprouts, seeds, relishes and drugs—are stronger than cultivated, irrigated herbs, because too much water weakens their power. Similarly, the venom of snakes that live in ponds and rivers is harmless, whilst mountain- or sand-dwelling snakes are lethal. §51

The same goes for fish. Those that are frail-bodied, thin-skinned, easy to chew, and that live in fresh, running waters are lighter (to digest) than those that live in lakes and ponds or are large and hard (to chew). On the whole, fish is cold and moist. For ill people, the most useful kind of fish lives in water that flows over rocks and sand; and fresh, roasted fish increases (the desire for) sexual intercourse, especially when (that fish) is fat. §52

All juvenile animals, plants and fruits, from the moment they come forth, tend to be cold and viscid, and make for bad nutriment—because of that coldness, neither their heat nor their moisture are (yet) established; once old, large and hard, they (also make for) bad nutriment, because (then) watery (stuff) has gained the upperhand; (only) when they are balanced in terms of age and size, and (taken at the right) time (of year), do they (provide) good, light, delicate nutriment. The most laudable flesh-meat is (obtained from) kids and lambs; the toughest is that of buffalos, cows, camels, stags, and other such dwellers of plains and mountains. The best organ-meat one can eat is the liver, because it generates excellent blood; the spleen generates black bile; the stomach is sinewy and slow to digest; and §53

بطيء الاستمراء والكلية أيضا رديئة فأما النيء من اللحم فإنه يولد دما رديئا غليظا إلا فيمن كان كثير التعب قليل البلة كثير النوم بعد الأكل، ورأيت الأمم الموصوفة بالشجاعة والسَّبُعية كالعرب والترك والديلم تستعمل النيء من اللحمان وما قرب من النيء فيورثهم ذلك السبعية والنجدة، وبلغني أن ملكا من ملوك الحبشان لا يؤثر عليه غيره ويقوى به على الجماع قوة عجيبة.

في الألبان

§54 وتختلف قوى ألبان الدواب وبيض الطير على قدر قوة الدواب والطير وعلى قدر الرِعْى والماء وأزمان السنة وسن الدابة فإن لبن الدابة الفتية الشابة الصحيحة القليلة الكد التي ترعى في الأسقاء وتحت الظلال هو أرطب وكذلك ما حُلِبَ أيام الصيف من دابة ترعى في المراعي الجبلية الأعذاء ويكون علفها حارا وإذا كانت الدابة كثيرة الكد والعمل فلبنها أحر وأيبس، والألبان في الجملة حارة رطبة وفيها ثلاث قوى: دسومة حارة رطبة ترطب البدن ومائية رقيقة باردة تبرد وغِلَظ أرضي يغذو البدن، وأغلظ الألبان ألبان الجواميس والبقر وألطفها ألبان النساء ثم ألبان الأتن.

في الأشربة

§55 فأما الأشربة فما كان من عنب جبلي عذي فإنه أيبس من السهلي فأما الجبلي العذي فإنه ينفع المشايخ وأصحاب البرد والرطوبات والسهلي السِقْى ينفع الشباب والمحرورين،

---

١ فأما النيء : فاما التى Ox, واما التى Is || رديئا Ox : – Is ٢ الأكل Ox : الطعام Is || السبعية Ox : السعه Is ٣ النيء[1] : التى Ox, السى Is || السبعية Ox : السعه Is ٤ ملوك الحبشان Ox : الحبسان Is || يقوى Ox : فمموى Is ٥ في الألبان Ox : – Is ٦ قدر[1] Ox : – Is || و[4] Ox : – Is ٧ الشابة Ox : – Is || كذلك : كان لذلك Is ٧–٨ الصحيحة القليلة الكد فيه Ox : – Is ٨ في الأسقاء و Ox : – Is || هو Ox : هى Is || فهو ارطب من لبن Ox : من Is || الأعذاء Ox : – Is ٩ فلبنها Ox : فهو Is ١٠ فيها ثلاث Ox : – Is || رطبة Ox : – Is ١١ أرضي Ox : ارض و Is || ألبان[1] Ox : – Is || ألبان النساء : البان الشآ Ox, النسا Is ١٢ في الأشربة Ox : – Is ١٣ عذي : غذي Ox, غدى Is || السهلي Ox : السهل Is || العذي : الغدي Ox, الغدى Is ١٤ البرد Ox : السرود Is || السهلي Ox : السهل Is || و[4] Ox : – Is

the kidney is bad, too. As for raw meat, it generates bad, thick blood, except for someone who does a lot of labouring, (holds) little moisture, and gets plenty of sleep after food. I have seen peoples who are famous for (their) bravery and ferocity—like the Arabs, the Turks, or the Daylamites[57]—and who regularly (eat) raw or almost raw meats, which is precisely what makes them brave and ferocious. And I was told that one of the Ethiopian kings liked nothing better (than raw meat), and that it gave him astonishing sexual power.

On milks

The milks of (land) animals, as well as (the whites) of bird eggs, have differ- § 54
ent powers, in proportion to the power of that animal or bird. (These powers moreover) depend on the pasture, on the water, on the time of year, and on the age of the animal—thus, the milk of relatively young, healthy beasts that labour little and graze in the shade on irrigated lands, is very rich in moisture; the same (is true) for beasts that are milked during the days of summer, that graze on remote, hilly pastureland, and whose forage is warm. If (however) the beast labours and toils a lot, then its milk is hotter and drier. On the whole, milk is hot and moist, and has three (basic) powers: a hot, moist fattiness which humidifies the body; a cold, thin wateriness which cools (the body); and an earthy thickness which nourishes the body. The toughest milk is that of buffalos and cows, the tenderest that of women, followed by ass's milk.

On wines

Concerning wines: what is (made) from wild mountain grapes is drier than § 55
(what is made) from lowland (grapes); as for wild mountain (grapes), they benefit old men and those who possess (a lot of) coldness and moistures, whilst irrigated lowland (grapes) benefit young people and those who are

57 The Daylamites were a people who, from as early as the 1st century BCE, inhabited the mountainous regions of northern Iran, along the southwestern shore of the Caspian Sea; they successfully resisted early Arab efforts to conquer their land, remained pagans until well into the 10th century CE, and generally had the reputation of being the most barbarous and odious enemies of Islam; employed as mercenaries already by the Sasanian kings, they continued to play a similar role in the armies of the ʿAbbāsid caliphs; in Ṭabarī's days, the caliphs had also begun to recruit Daylamites, alongside Turks, for the Royal Guard—see FMDey *passim*.

وكلما عتق فإنه يزداد حرارة ولطافة وينفع من الفضول الباردة الغليظة ويفرح القلب ويحسن اللون ويطلق تعقد اللسان ويشجع الجبان ويشوّق إلى كل متوَّق بَهيج شهي ويذكّر بالأوانس والأزواج وكلما اشتدت حمرته وغلظ كان أكثر توليدا للدم، وما كان منها عفِصا حامضا فهو أقل دما وغذاء وما كان منه حلوا فهو يقرقر وينفخ ويولد سددا في البدن.

في الأدهان وغيرها

§56 وكل دهن مثل قوة الشيء الذي يطيَّب به مثل الزنبق الذي هو حار كالياسمين ودهن البنفسج الذي هو بارد والخيري والسوسن اللذان هما معتدلان بينهما، وكل كامَخ فإن قوته مثل قوة الشيء الذي عُمِلَ منه لِما فيه من اللين والتعفين، فأما المري فإنه معتدل في الحر واليبس.

§57 والخل يذكر جالينوس أن فيه حرارة لطيفة وأنه يعمل في الأجسام ما تعمل النار، وكما أن النار تغيِّر ما أُلْقِيَ فيها في زمان قصير فإن الخل يغيِّر الحديد والصخر وغيرهما في زمان أطول من ذلك.

§58 وقوة كل أفشرج ورب مثل قوة الشيء الذي يُعتصر منه، فرب الرمان والتفاح والحصرم والإجاص بارد يبرد الحرارات ويسكن العطش ورب الآس بارد يدبغ المعدة ويحبس البطن وكذلك رب السفرجل معتدل يدبغ المعدة ويقويها ويحبس البطن، والشواريز ثقيلة ملطخة للمعدة مورثة للسدد.

---

١ كلما : كل ما Ox, Is || اللون : السدن Is ٢ تعقد Ox : – Is || الجبان : الحنان Is || متوق بهيج : موىى بهىح سمى Is ٣ كان Ox : + اقوى و Is || توليدا للدم : Ox بولىد الدم Is || عفصا : Is عفضا Ox ٤ سددا في البدن : Ox السدد فى البطن Is ٥ في الأدهان وغيرها Ox : – Is ٦ كل شهي : Ox دهن : Ox كلما كان دهنا فوته Is || الزنبق : Ox الزمو Is || كالياسمين : Ox مثل الياسمس Is ٧ بارد + : Ox كالىمسح Is || اللذان : Is اللذين Ox || فإن Ox : – Is ٨ اللين : اللبَن Ox, اللىن Is || معتدل + : Ox فى الجوف Is ٩ وكما : Ox فكلما Is || تغير + : Ox كل Is ١١ و¹ Ox : – Is || أفشرج : افسرح Ox, افسرح Is ١٢-١٣ معتدل ... يدبغ Ox : – Is ١٣ الشواريز : Is السوارير Ox

hot-tempered. The older (the wine) is, the hotter and milder it becomes—(such a wine) is useful against cold, tough residues, it gladdens the heart, embellishes the complexion, loosens the inhibited tongue, emboldens the coward, arouses a yearning desire for all that is bright and beautiful, and evokes the lovers and the spouses. The redder and denser (the wine) is, the more blood it generates; pungent, sour (wines give) less blood and nutriment; sweet (wine causes) rumbling and bloating, and produces obstructions in the body.

On oils and others

Any (given) oil shares the power of the thing that was used to perfume it—for example, jasmine oil, which is hot like the jasmine (plant); sweet violet oil, which is cold; or gillyflower and lily (oil), which are both balanced in between. (Similarly), the power of any vinegar-based pickle corresponds to the power of the thing with which it was made—this is due to the suppleness and fermentiveness (of vinegar). As for garum, it is balanced between heat and dryness. §56

Galen mentioned that vinegar possesses a subversive heat, and that it acts upon matters in analogy to fire: just as fire, in a short period of time, changes everything that is thrown into it, so does vinegar, (albeit) in a longer period of time, change (even) iron, rock, and other such (stuff).[58] §57

The power of any press-juice or syrup corresponds to the power of the thing from which it was extracted—thus, the syrup (obtained from) pomegranates, apples, unripe grapes, or plums is cold, tempers fever heat, and quenches thirst; the syrup of myrtle (berries) is (also) cold, fortifies the stomach, and tightens the belly; similarly, the syrup of quinces, (which are) balanced, fortifies and invigorates the stomach, and tightens the belly. Conserves are heavy; they soil the stomach, and leave obstructions. §58

58 Galen, considering vinegar (*ὄξος*) to possess an 'obscure heat' (*θερμότης ἀμυδρά*), further notes: "Id quod in primis efficere acetum valet, quippe quod secet atque erodat non modo animantium corpora, sed et grumos, callos, lapides, fictile, aes, ferrum, plumbum; ac tantum non velut ignis cuncta pervadat ac penetret, nec ulla possit substantia penetrationis ejus vehementiam aut sustinere, aut illi reluctari ac resistere", see GalKü 11/420 and 425 (*Περὶ κράσεως καὶ δυνάμεως τῶν ἁπλῶν φαρμάκων*, cf. FiCG no. 78).

في الأغذية اللطيفة والغليظة

§59 إن من الأغذية أغذية لطيفة مثل الشفنين والدراج والعصافير الأهلية والعنب والتين وقصب السكر والزبيب المنزوع الحب والخمر البيضاء لطيفة وماء الشعير والسمك الذي يكون في أنهار عذبة صخرية، وقالوا: إن السمك المالح أيضا يحلل البلغم ويلطفه، وكذلك الحبق والزعتر والثوم والبصل والشبث والحرف والكمون والكراث والجرجير والفجل وكذلك كل طيب الريح حِرِّيف، وهذه البقول التي ذكرتها إذا كانت رطبة فهي أغذية وإذا جفت صارت أدوية.

§60 ومنها ما يورث الغلظ مثل لحوم الأيائل والمعز الكبار والإبل العوامل والأرانب والأعيار والتيوس والسمك الذي يغلظ قشوره والأكباد والكلى والمخاخ والضروع والألسنة والأجبان والتمر واللفت والفطر ولحم الأترج والخمر الحلو والرب، فأما الأغذية المعتدلة فهي مثل الفراريج والديكة والورشان والحمام وفراخ العصافير والهندباء والخس والهليون.

في قوى المزاجات وعلل هيجانها وأسباب توليدها والدلائل الواضحة على ذلك كله

§61 قد قدّمتُ ما رأيت تقديمه بإيجاز من قول وصواب من تدبير البدن على ما قالت الحكماء، وأخّرتُ ذكر الطبائع والمزاجات لئلا تنقطع المواد ويتقارب المعنى الذي إليه قصدتُ وإياه أردتُ.

٢ الشفنين و Ox : – Is ‖ العنب والتين Ox : التين والعنب Is ٤ أنهار عذبة Ox : الانهار العذبه Is ‖ يحلل :
٨ Is – : Ox إذا[1] ٦ Is السعـز : Ox الزعتر ‖ Is سلقطه ومـل : Ox يلطفه وكذلك ‖ Is سملل ,Ox محلل
٩ Is – : Ox والإبل العوامل ‖ Is والسـران + : Ox المعز ‖ Is لحم + : Ox و[2] ‖ Ox الايابيل : Is الأيائل
قوى ١٢ Ox والفراخ + : Is الحمام ١١ Is – : Ox فهى ‖ Is القطر : Ox الفطر ١٠ Is – : Ox والأجبان
Is – : Ox من قول وصواب ‖ (بإنجاز fortasse) Ox, Is باـحاز : بإيجاز ‖ Is – : Ox قد ١٤ Is قوه : Ox
Is التى : Ox الذي إليه ‖ Is – : Ox المواد ‖ Is الصنايع : Ox الطبائع ‖ Is اجزات : Ox أخرت ١٥

On tender and on tough nourishments

Some nourishments are tender, such as turtledoves, francolins, domestic sparrows, grapes, figs, sugarcane, and seedless raisins; white wine (too) is tender; (further) barley water, and fish that live in clean, rocky rivers. (People) say that salt(water) fish also dissolves and mitigates phlegm. The same (is true) for basil, savory, garlic, onions, dill, gardencress, cumin, leek, rocket, radish, and all that has a fragrant smell and a pungent (taste). These plants which I just mentioned are (items of) food when they are fresh, but become drugs when they are dry. § 59

Some (nourishments) are hard to digest,[59] such as the meats of stags, fully grown goats, work-camels, hares, onagers, billy goats, and fish that have thick scales; (further), livers, kidneys, marrows, udders, tongues, cheeses, dates, turnips, mushrooms, citron pulp, sweet wine, and (many) syrups. As for balanced nourishments, they (include items) such as chicks, cocks, ringdoves, pigeons, young sparrows, wild chicory, lettuce, and asparagus. § 60

On the powers of (humoral) mixtures, the reasons for their agitation, the causes of their generation, and clear evidence for all that

I have now introduced, in a few words and (with a focus on) bodily regimen, what I considered worthy of introduction and pursuant to the teachings of the sages. I have (so far) delayed any talk about temperaments and mixtures in order to avoid a disruption of arguments, and to ensure a gradual approach to the core of what I wish to say. § 61

---

59 The phrase *wa-minhā mā yūriṯ al-ġilaẓ*, which is attested as such by both manuscripts, literally means “and some (nourishments) bequeath coarseness”; there is a vague textual parallel in the *Paradise of Wisdom*, reading *wa-minhā mā ṣāra ġalīẓ*^an^ *li-ṣalāba fīhī* “and some (nourishments), due to their hardness, become coarse (when digested)”, see ṬabFir 119,5 f. My translation is conjectural in that sense.

§62 فالطبائع العظمى أربع: الحرارة والبرودة والرطوبة واليبوسة، واثنتان منها فاعلتان بإذن الله تعالى وهما الحرارة والبرودة واثنتان مفعولتان وهما الرطوبة واليبوسة، وقوام كل شيء في العالم بهذه الأربعة لا يستغني عن هذه الطبائع شيء من حيوان ولا نبات ولا يقوم إلا بها أعني بالأرض والماء والهواء والنار، ومن هذه تتولد المزاجات الأربعة أعني الدم والبلغم والمرة الصفراء والمرة السوداء ومن هذه المزاجات تتكون أبدان الناس وسائر الحيوان.

§63 فأما الدم فحار رطب حلو يشبه بالهواء في قوته وحركته وبالربيع وجهة التّيْمَن وريحها الجَنوب وبسن الصبى، ومن الدم يكون الفرح والنشاط والمزح والبِشْر وطيب العِشْرة والضحك وإشراق الوجه والقوة على الجماع ومنه يكون الجدري الدموي والحمرة الدموية وبعض الآكلة وبعض الرمد والحمى الدائمة الحارة، ويكون تولُّد الدم من الأغذية فما كان منها حارا رطبا فهو يصير دما وما كان منها من نارية فإنه يكون منه الصفراء وما كان منها من مائية فإنه يكون منه البلغم وما كان منها من أرضية فإنه يكون منه المرة السوداء وما زاد على ذلك في البرد واليبس صار منه العظام وما كان فيها من دسومة فإنه يكون منه الأدمغة والمخاخ وما كان فيها من رطوبة مع أدنى صلابة صار عصبا وعضلات، فهذا في تولُّد الدم من الأغذية وتصوُّر الأعضاء من الدم.

§64 فأما المرة الصفراء فحارة يابسة مرة من جوهر النار في قوتها وحركتها وأمراضها وأمراض المرة حارة يابسة وهى تشبه بالصيف وسن الشباب وبجهة المشرق وريحها الصَبا والقَبول، ومنها النزق

---

١ أربع : Ox اربعه Is ‖ اثنتان : Ox اسـان Is ‖ فاعلتان : Is فاعلة Ox ‖ تعالى : Ox – Is ٢ اثنتان Ox
وهما : Ox – Is ٣ الأربعة : Ox الطبايع الاربع Is ‖ عن هذه الطبائع : Ox عنها Is ٤ اسـان :
المزاجات الأربعة : Ox الاربع Is ‖ المرة[2] : Is – Ox ٥ تتكون : Ox فـكون Is ‖ الحيوان : Ox الحيوانات
Is ٦ فأما الدم : Ox فـالدم Is ‖ التيمن : Ox الـسمن Is ‖ ريحها : Ox + التى هى Is ٧ والمزح : Ox –
Is ‖ العشرة : Ox + والسرور Is ‖ إشراق : Is اشراف Ox ٨ الدموي : Ox الدمى Is ‖ الدموية : Ox –
Is ‖ الرمد : Ox الزهد Is ٩ ويكون : Ox فـكون Is ‖ منها : Ox – Is ‖ فهو : Ox هو اشـا Is ‖ دما Ox
: + وحما Is ‖ وما : Is فما Ox ٩–١٠ الصفراء وما ... كان منها من نارية : Ox – Is ١٠ منها[2] : Is فيها Ox ‖
منه[2] : Ox من Is ١١ من : Ox – Is ‖ منه[1] : Ox من Is ‖ العظام : Ox الطعام Is ١٢ فيها[1] : Ox – Is ‖
منه : Ox مـها Is ‖ فيها[2] : Ox مـها Is ١٣ فهذا : Ox فهده Is ١٤ المرة[1] : Ox – Is ‖ فحارة : Ox فحره
Is ‖ مرة : Ox – Is ‖ في : Ox و Is ١٥ حارة : Is حار Ox ‖ بجهة : Ox بهحه Is ‖ ريحها : Ox + التى هى
Is ١٥–٩٠.١ النزق والحدة والخفة : النزق والحدة والخفة Ox, البرق والحده والحفه والـسرق Is

There are four principal temperaments: heat, coldness, moisture, and dryness. Two of them are active—with the permission of God the Sublime—, namely heat and coldness; two of them are passive, namely moisture and dryness. The coherence of every thing in this world depends on these four temperaments—no animal, no plant can exist without them, nor survive without their (elemental counterparts), namely earth, water, air, and fire. From these (elements) are generated the four (basic) mixtures, namely blood, phlegm, yellow bile, and black bile, and from them (in turn) are formed the bodies of humans and all other creatures.[60] § 62

As regards blood, it is hot, moist and sweet; it is similar to air in terms of its power and its movement, and (associated) with spring, with the direction of the south and its wind the *ǧanūb*, and with the age of childhood. From blood arise joyfulness, liveliness, wittiness, cheerfulness, chumminess, laughter, a radiant face, and the power to (perform) sexual intercourse; yet it (also) causes bleeding smallpox, bleeding erysipelas, some gangrenous sores, some eye inflammations, and chronic high fever. Blood is generated from nourishment: hot, moist (nourishment) becomes blood, fiery (nourishment) produces yellow bile, watery (nourishment) produces phlegm, and earthy (nourishment) produces black bile; leftover coldness and dryness turn into bone, (leftover) fattiness turns into brain and marrow, and (leftover) gooey moisture turns into nerve and muscle. So much for the generation of blood from nourishment, and the formation of the organs from blood. § 63

As regards yellow bile, it is hot, dry and bitter; it is essentially like fire in terms of its power, its movement, and its diseases, which latter are (also) hot and dry; it is associated with summer, with the age of youth, and with the direction of the east and its wind the *ṣabā* or *qabūl*. From (yellow bile arise) § 64

60 For elements, temperaments and 'mixtures' (scil. humours) see UllIM 56–60 (an English summary of al-Maǧūsī's [d. late 4th/10th century] expositions on that subject); cf. also the diagrammatic representations in §§ 67 and 68.

والحدة والخفة والتطاول والعُجْب ولون من غلبت عليه الصفرة ومعها تكون درابة اللسان والتوقد وشهوة الجماع على أن زرع أصحابها أقل من زرع أصحاب الدم ومنها تكون الوسوسة والهذيان وحمى الغب وبعض البرسام الحار واليرقان في الصفراء.

§65 وأما البلغم فبارد رطب مالح يشبه بالماء في قوته وحركته وبالشتاء وبسن الشيخوخة وبجهة المغرب وريحها الدَبور، ولون من غلب عليه البلغم البياض ومنه يكون الثقل والبلادة واللين والمواتاة وفتور شهوة الجماع ورطوبة الزرع ومنه يكون الفالج واللقوة والنسيان وحمى كل يوم وقروح في الوجه ألوانها إلى البياض والتقطير والبواسير واسترخاء الأعضاء وبعض الاستسقاء والبرص والارتعاش وبعض الصرع.

§66 وأما السوداء فباردة يابسة حامضة وبحموضتها تحرك شهوة الطعام وحركتها وقوتها حركة أرضية إلى أسفل وهي تشبه بالأرض والخريف وبسن الاكتهال وبجهة الجُدَيّ وريحها الشِمال، ومنها يكون طول الصمت وشدة الحقد ولطافة الفكر، ويقال: إنها كانت الغالبة على أكثر الفلاسفة المتقدمين، ولصاحبها سورات مثل سورات السباع وذلك بعد طول صمت وبجَزَع وصبر ومنها تكون حمى الربع والفزع والصرع والسرطان الذي هو الداء العياء وأمراضها متطاولة مزمنة لا يكاد يبرأ صاحبها مثل المرض الكاهني.

---

١ غلبت Ox : علب Is ٢ زرع^{2} Ox : – Is || تكون Ox : يكون Is ٣ اليرقان Is : + و Ox ٤ بالماء Ox : الما Is || وبجهة Ox : بسن الشيخوخة وبهجه Is : سن السـوح وبهجه ٥ ريحها Ox : + التي هي Is || لون Is : كون Ox ٦ منه Ox : منها Is || اللقوة Ox : القوه Is || وحمى Ox : – Is || في الوجه Ox : وجه Is ٧ استرخاء Ox الاسترخاء : Is || الارتعاش Ox : الرعاس Is ٩ تحرك Ox : يحرك Is ٩–١٠ حركة أرضية إلى أسفل Ox : داحل الى الارض بحركه سبه Is ١٠ بسن Ox : سن Is || بجهة الجدى Ox : بهحه الحرسا Is || ريحها Ox : + التي هي Is ١١ طول Ox : – Is || الفلاسفة Is : + و Ox ١٢ ذلك Is : لذلك Ox ١٣ الربع Ox : ربع Is || متطاولة Ox : مطاوله Is ١٤ صاحبها Ox : – Is

---

61 The reading *nazaq* “capriciousness” is in line with ṬabFir 42,14; the Oxford manuscript reads, nonsensically, برق and the Istanbul manuscript leaves the lemma undotted to read برق.

capriciousness,[61] irascibility, frivolity, arrogance, and vanity; the complexion of choleric types is yellow, and they are nimble-tongued, fervid, and keen on sexual intercourse, although they possess less semen than the sanguine types; yellow bile is responsible for hearing voices, hallucinations, tertian fever, some (forms of) hot phrenitis,[62] and jaundice.

As regards phlegm, it is cold, moist and salty; it is similar to water in terms of its power and its movement, and (associated) with winter, with old age, and with the direction of the west and its wind the *dabūr*. The complexion of the phlegmatic type is white; (phlegm) is responsible for heaviness, dullness, meekness, affability, a lukewarm interest in sexual intercourse, and moisture-saturated semen; it (further) causes hemiplegia, facial paralysis, forgetfulness, quotidian fever, whitish ulcers of the face, dribbling (urine), haemorrhoids, laxity of the organs, some (forms of) dropsy, white leprosy, tremor, and some (forms of) epilepsy. § 65

As regards black bile, it is cold, dry and sour, and through that sourness stirs an appetite for food; its movement and its power are groundwards and earthlike; it is (associated with) autumn, with middle age, and with the direction of the pole star and its wind the *šimāl*. From (black bile) arise taciturnity, disaffection, and intellectual refinement. It is said: (black bile) was dominant in most ancient philosophers. The melancholic type is prone to violent outbursts in the manner of predatory animals, especially after long periods of silence, but (he is also known for his) courage and patience; (black bile) causes quartan fever, (states of) anxiety, epilepsy, and cancer, which is an incurable condition; the illnesses (related to black bile) are lengthy and (often) chronic, and hardly ever does their victim (fully) recover—witness the (so-called) diviner's disease.[63] § 66

62 On "phrenitis" (*barsām*) cf. note 51 above.

63 *al-maraḍ al-kāhinī* "diviner's disease" (calque of *ἱερὰ νοῦσος* "holy disease" [i.q. *πάθος ἱερόν*]) is a folkmedical designation of epilepsy, see ThErk 29 f., WkaS 1/418b and UllMed 235; in the *Paradise of Wisdom*, Ṭabarī elaborates: "Epilepsy (*ṣarʿ*)—that is *ifīlibsiyā* [< *ἐπιληψία*]—is called diviner's disease by the people because some of these (victims) can predict the future (*yatakahhan*), and strange things appear to them", see ṬabFir 138,6 f. The idea expressed earlier on in this paragraph, namely that black bile "was dominant in most ancient philosophers", is Aristotelian, as has been pointed out most recently by Peter Pormann, who translates the relevant passage in the *Problemata* (30.1) as follows: "Why is it that all those men who excel in philosophy, politics, or the arts appear to be melancholics?", see PorMel 7.

§67 وهذه دائرة صوّرتها الحكماء على شكل الدنيا وأظهروا العيان عليها، إن الله تبارك وتعالى بلطف تدبيره جعل تمازج هذه الطبائع الأربع من حواشيها وأطرافها التي تشاكل بعضها بعضا كامتزاج الحر بالبرد والبرد بالحر على ما في هذا الشكل حتى خلق منها هذا العالم مع ما بينها من التعادي والتضاد:

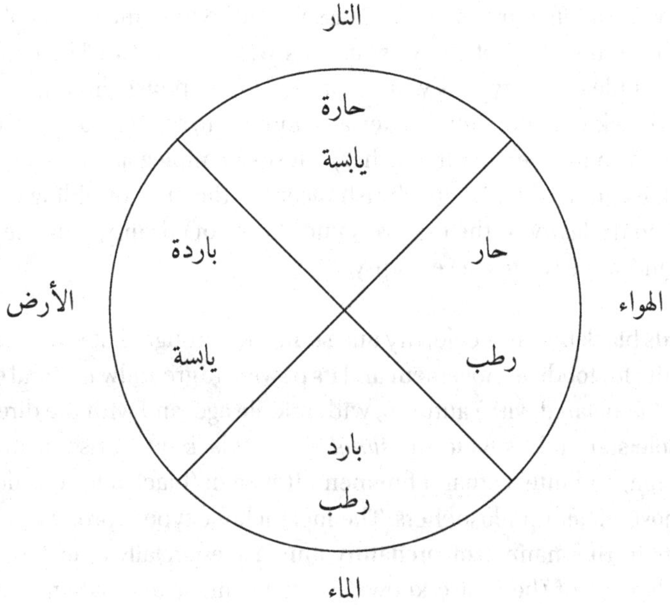

اطرافها وحواسـها : Ox حواشيها وأطرافها التي تشاكل ٢ Is – : Ox تبارك وتعالى || Is – : Ox عليها ١
الباس : Ox التضاد ٤ Ox بينهما : Is بينها || Ox بالبرد : Is بالحر || Ox بـالحر : Is بالبرد ٣ Is لـمشاكل
: حارة ٦ The following illustration is missing from the Oxford manuscript. | Is والـحارب
Is حار

Here is a circle which the sages have drawn in the shape of the earth, in order to illustrate the evidence. So did God the Blessed and Sublime, in His benevolent scheme, arrange for the four temperaments to intermix at the seams and fringes which are woven into each other—like the blending of heat with coldness or coldness with heat—, as can be seen from this figure; and thus He created from them the world, despite their mutual hostility and opposition:[64] §67

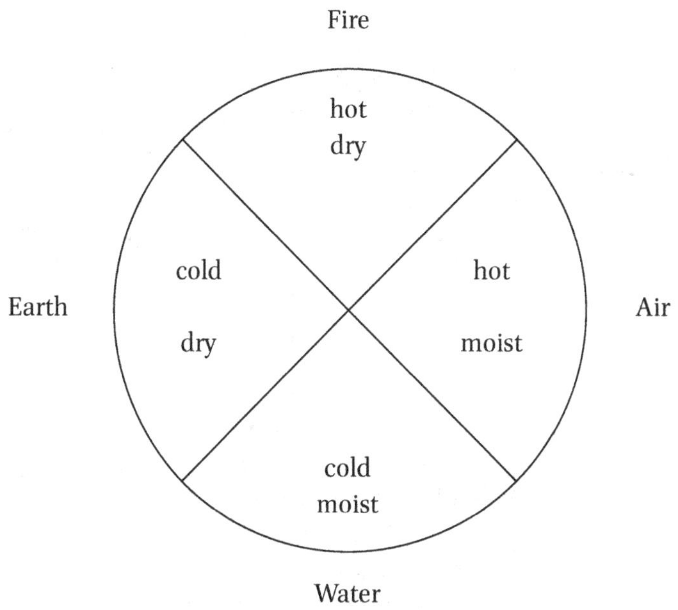

64 The following illustration is missing from the Oxford manuscript.

§68 وهذا شكل صليبي في كل جانب منه أشياء متشابهة القوى والأفعال عجيب جدا:

جميع ما في هذا الربع حار يابس: النار، الصفراء، الصيف، الشباب، المشرق وريحه وهى الصبا، الساعة الرابعة والخامسة والسادسة من النهار، ومن قوى البدن القوة النفسانية والحيوانية والجاذبة، ومن المذاقات المرارة، ومن الكواكب المريخ والشمس، ومن البروج السرطان والأسد والسنبلة

جميع ما في هذا الربع حار رطب: الهواء، الدم، الربيع، الصبي، التيمن وريحه وهى الجنوب، الساعة الأولى والثانية والثالثة، ومن القوى القوة الطبيعية والهاضمة، ومن المذاقات الحلاوة، ومن الكواكب المشتري والعطارد، ومن البروج الحمل والثور والجوزاء

الوسط

جميع ما في هذا الربع بارد يابس: الأرض، السوداء، الخريف، الكهولة، الجربياء وهى الشمال، الساعة السابعة والثامنة والتاسعة، ومن القوى القوة الماسكة، ومن المذاقات الحموضة، ومن الكواكب زحل، ومن البروج الميزان والعقرب والقوس

جميع ما في هذا الربع بارد رطب: الماء، البلغم، الشتاء، الكبر، المغرب وريحه وهى الدبور، الساعة العاشرة والحادية عشر والثانية عشر، ومن القوى القوة الدافعة، ومن المذاقات ما أشبه منها الملح، ومن الكواكب القمر والزهرة، ومن البروج الجدى والدلو والحوت

---

١ وهذا شكل ... عجيب جدا Ox : – Is | Variant documentation for the following diagram has been kept to a minimum. ٢ الربع Ox : السربع Is (et seqq.) ٧ الحيوانية والجاذبة Ox : الحملسه الحيوانه والهاضمه Is ٨ المذاقات : المراحات Ox, المداوات Is ١٢–١٣ التيمن وريحه وهى الجنوب Is : الجنوب وريحها Ox ١٢ الكهولة Is : الاكتهال Ox ١٣ الجربياء وهى Is : – Ox ١٤ الوسط Ox : – Is ١٦ الحموضة Ox : العفص الحماص Is ١٦–١٧ المشتري والعطارد Ox, Is : القمر والزهرة ṬabFir 59 ٢٠ الشتاء الكبر Is : Ox non legitur ٢٣ الدافعة Ox : الهاضمه Is ٢٤ ومن المذاقات ما أشبه منها المل[ح] Is : Ox non legitur ٢٥ ومن الكواكب القمر والزهرة ṬabFir 59 : – Ox, Is ٢٧ This diagram marks the end of the Istanbul manuscript.

And here is a rather marvellous cross-shaped figure, each side of which contains things that are associated with the same powers and effects: §68

	Everything in this quarter is hot and dry: fire; yellow bile; summer; youth; the east and its wind the *ṣabā*; the fourth, fifth and sixth hour of the day; among the faculties of the body,[65] the psychical, the animal and the attractive faculty; among the tastes, bitterness; among the planets, Mars and the Sun; among the signs of the zodiac, Cancer, Leo and Virgo[66]		
Everything in this quarter is cold and dry: earth; black bile; autumn; middle age; the north wind which is the *šimāl*; the seventh, eighth and ninth hour; among the faculties, the retentive faculty; among the tastes, sourness; among the planets, Saturn; among the signs of the zodiac, Libra, Scorpio and Sagittarius	The Middle	Everything in this quarter is hot and moist: air; blood; spring; childhood; the south and its wind the *ǧanūb*; the first, second and third hour; among the faculties, the natural and the digestive faculty; among the tastes, sweetness; among the planets, Jupiter and Mercury; among the signs of the zodiac, Aries, Taurus and Gemini	
	Everything in this quarter is cold and moist: water; phlegm; winter; old age; the west and its wind the *dabūr*; the tenth, eleventh and twelfth hour; among the faculties, the excretory faculty; among the tastes, what resembles that of salt; among the planets, the Moon and Venus; among the signs of the zodiac, Capricorn, Aquarius and Pisces[67]		

65 On the concept of faculties (*quwan*) cf. note 40 above.

66 On astrological components of early Islamic medicine see e.g. UllMed 254ff.; further the two tables in UllNGw 348 and 350. Ṭabarī's keen interest in iatromathematics is most clearly reflected in an essay on the foundations of astrology, which forms part of his medical compendium *Paradise of Wisdom* (ṬabFir 541–556).

67 This diagram marks the end of the Istanbul manuscript.

في علل هيجان الطبائع والمزاجات وغلبتها

§69 يكون هيجان الحرارة إما من تعب شديد أو من مطعم حار مثل الخردل والثوم والفلفل والسذاب والعسل وإما من كثرة اللبث في الشمس وإما من عفونات في البدن فتلتهب فيه الحرارة كما تلتهب في الزبل إذا عفن وإما من انسداد مجاري البدن واحتباس الحرارة فيها كما تُحتبس النار في الأتاتين وإما من سخافة البدن وانخزاله لأنه يلبّس البدن حينئذ وتهيج الحرارة وإما من مقاربة أجساد حارة وإما من كثرة الحركة والسهر والصوم الدائم ومن الجماع أو من هيجان الدم أو من قلة إسهال البطن.

§70 ويكون هيجان البرد إما من كثرة الراحة أو من أطعمة وأشربة باردة مثل الجبن واللبن والسمك وماء الثلج والجليد وإما من برد الهواء وإما من مقاربة أجساد باردة، وإن أفرط مفرطٌ في الجماع أو السهر أو الفكر أو إخراج دم أو إسهال بطن حتى تجاوز الحد فيه بَرُدَ الجسم وانطفأت حرارته.

§71 ويكون هيجان اليبس إما من تعب شديد وإما من قلة الطُعَم أو أطعمة يابسة أو مقاربة أجساد يابسة ومن كثرة الاستحمام بمياه مالحة أو كبريتية أو طول السهر أو كثرة الهم والفِكَر الرديئة.

§72 ويكون هيجان الرطوبات إما من طول الدعة وأطعمة وأشربة رطبة ومقاربة أجسام رطبة أو كثرة الاستحمام بمياه عذبة أو كثرة النوم بعد الطعام.

في الدليل على مزاجات الأبدان والغالب عليها

§73 إن الرؤية والألوان تدل على أشياء كثيرة باطنة، كما يدل سقوط شعر الحاجب وقروح الوجه وتشنج الأظافير واستدارة العين على فساد الرئة والجذام وكما تدل صفرة الوجه والعين على اليرقان وبياض الوجه والشفة على البواسير، وقد يُستدل بالاعتبار والنظر ليس على ذلك فقط بل على الشِيَم المستورة.

---

٤ الزبل : الذيل Ox || الأتاتين : الاتانين Ox ٨ مقاربة : مقارنة Ox ٩ انطفأت : انطفت Ox ١٠ مقاربة : مقارنة Ox ١٢ مقاربة : مقارنة Ox ١٥ باطنة : باطلة Ox

On the reasons for the agitation and predominance of temperaments and mixtures

Heat may be agitated by hard labour; by hot food, like mustard, garlic, pepper, rue, or honey; by frequent exposure to the sun; by putrefying matters inside the body, which unlock heat as they do in rotting dung; by a constriction of bodily channels with detainment of heat, just like fire is detained in kilns; by physical frailty and torpidity, in which case the body becomes confused and the temperature rises; by being close to hot objects; by too much exercise, or sleeplessness, or constant fasting, or (immoderate) sexual intercourse; by an agitation of blood; or by infrequent purging of the belly. § 69

Coldness may be agitated by too much rest; by cold foods and drinks, like cheese, milk, fish, or snow water and ice (water); by cold air; or by being close to cold objects; (likewise), the body cools off and its heat dies down if somebody exceeds the (proper) bounds of sexual intercourse, waking, thinking, bloodletting, or purging the belly. § 70

Dryness may be agitated by hard labour; by infrequent meals; by dry foods; by being close to dry objects; by overmuch bathing with saline or sulphurous water; by prolonged sleeplessness; or by worrying a lot and having bad thoughts. § 71

Moistures may be agitated by prolonged idleness; by moist foods and drinks; by being close to moist objects; by overmuch bathing with fresh water; or by oversleeping after meals. § 72

On how to discern (specific) mixtures and their preponderance over the body

The external aspect and the complexion (of a person) can give clues to a variety of internal conditions—for example, hairloss on the brows, ulcers of the face, rugged fingernails and globular eyes indicate a corruption of the lungs or (early) leprosy; and whilst yellowness of the face and eyes indicates jaundice, whiteness of the face and lips indicates haemorrhoids. Sometimes, reflection and observation are guides not only to such (underlying illnesses), but also to hidden characteristics. § 73

§74 فقد قالت الحكماء: إن مَن عظم رأسه جدا دل على أنه جلف فدم ومَن صغر رأسه جدا دل على قلة الدماغ وفساد الذهن وإن كانت عينه غائرة صغيرة دلت على قلة الفهم وسوء التركيب، وإن كانت صورته ثورية دلت على انقياد وكد وغلظ وإن كانت صورته ثعلبية دلت على خبث وجريرة وزوغان وإن كانت صورته قردية دلت على طرب وقصف وفجور وإن كانت صورته أسدية دلت على شجاعة وكرم وإقدام وإن كانت صورته بعيرية دلت على خور وهوج وخفة وإن كانت صورته ديكية دلت على هراش وحراسة وشجاعة وسخاء وغيرة وكثرة جماع.

§75 فكذلك يُستدل على هيجان هذه المزاجات بعلامات قد ذكرها الحكماء: فمن الدليل على غلبة الدم أن من غلب عليه الدم كان جميل الوجه أصهب الشعر مخضب البدن كثير الضحك حريصا على الجماع واللهو مع حمرة الوجه والعين وحلاوة الفم وامتلاء العروق وحرارة مجسة البدن وكثرة النوم وأن يرى في منامه اللعب والملاهي والرياض والملابس الحسنة وشهوة الجماع وكثرة الإنعاظ وغزارة الزرع وكثرة شعر البدن.

§76 ومن الدليل على الصفراء وهيجانها أن من غلبت عليه الصفراء كان نزقا جريئا حديدا خفيفا نحيف البدن مصفر اللون أسود الشعر وأقل شعرا من صاحب الدم لأن الشيء الحار اليابس والبارد اليابس يضران بالشعر ويقلانه وذلك مع مرارة الفم ويبسه ويبس المنخرين وقلة النوم وسخونة البدن وشدة نبْض العروق وسرعة حرارة المجسة وكثرة الكلام والتشوق إلى برد الهواء وشدة الشبق وقلة الزرع وأن يرى في منامه النيران والحرب والمنازعات والشغب.

---

٤ صورته[1] : صورة Ox ١٣ مصفر : مضفّر Ox ١٤ يضران : نضران Ox ١٦ الشغب : السَّغَب Ox

---

68 Similar physiognomical observations, here attributed to certain "sages" (*ḥukamāʾ*), are found in the *Paradise of Wisdom*, where they are more specifically linked to the "physiognomists" (*ahl al-firāsa*) var. "the author of the *Physiognomy*" (*ṣāḥib al-Firāsa*), see ṬabFir 49,14–19 with note ٢. The latter is most probably a reference to Polemon of Laodicea's (fl. 2nd century CE) *Περὶ φυσιογνωμονίας* (Arabic title *Kitāb Iflīmūn fī l-Firāsa* [*wa-huwa kitābuhū *Fīsiyāġnūmunīqā**]), of which Ṭabarī seems to have extracted, here too, some small segments, perhaps proceeding from the (lost) Syriac translation of the text: for Polemon's corresponding observations on the size of the head see ScPhy 235,13 and 16f. = HoyPo 420,13 and 15f.; on similarities between human and animal features ScPhy 171–199 = HoyPo 384–394 (some correspondences); as regards Ṭabarī's short interpretation of "tiny hollow eyes", this does not correspond to any of Polemon's

The sages have said: a very large head suggests that the person is boor- §74
ish and dim-witted, a very small head suggests that he has little brain and an impaired mind; tiny hollow eyes suggest a lack of understanding and a bad (moral) constitution. A bull-like appearance suggests docility, tenacity and rudeness; a fox-like appearance suggests craftiness, wickedness and dishonesty; an ape-like appearance suggests excitability, revelry and debauchery; a lion-like appearance suggests bravery, nobility and audacity; a camel-like appearance suggests fragility, foolishness and frivolity; and a cock-like appearance suggests quarrelsomeness, vigilance, valiance, generosity, jealousy and a great sexual urge.[68]

From (other) signs, which the sages have (also) mentioned, one can draw §75
conclusions about the agitation of these mixtures. Thus, a predominance of blood in a person is indicated by (the fact that he has) a nice face, reddish hair, and an orange-brown body; he is full of laughter, and keen on sexual intercourse and amusement; moreover, his face and eyes are ruddy, he has a sweet taste in his mouth, his blood vessels are filled, and his body feels warm to the touch; he sleeps often, and in his sleep sees (places of) fun and entertainment, gardens and beautiful clothes; he has a (strong) sexual appetite, frequent erections, plenty of semen, and a lot of body hair.

An agitation of yellow bile is indicated by (the fact that) the person is §76
impetuous, reckless, abrupt, and frivolous; his body is slim, his complexion yellow, and his hair black, (although) he is less hirsute than the sanguine type—(this is so) because anything that is hot-dry or cold-dry damages the hair and thins it out; moreover, he has a bitter, stale taste in his mouth, and dry nostrils; he sleeps little, and his body is heated up; his arteries pulsate strongly and feel hot to the touch straightaway; he talks a lot, craves cool air, and is lustful but lacking in semen; in his sleep he sees fires and war, struggles and unrest.

---

lengthy declarations on the subject of ocular features (*firāsat al-ʿain*, see ScPhy 107–171 = HoyPo 340–384), nor does it occur as such in pseudo-Aristotle's *Φυσιογνωμονικά* (Arabic title *Kitāb Arisṭāṭālīs fī l-Firāsa*, see ScPhy 4–91 [Greek/Latin] = GhFir 3–50 [Arabic]), which may have been available to Ṭabarī in Ḥunain ibn Isḥāq's translation. Having said that, there is a striking and almost literal parallelism between Ṭabarī's expositions in §§73–74 and the beginning of the chapter on physiognomy (*firāsa*) in Masīḥ ad-Dimašqī's (d. after 225/840) long epistle written for the caliph Hārūn ar-Rašīd (reg. 170/786–193/809), see DimRH 429; also note that §73 has no corresponding passage in the *Paradise of Wisdom*.

§77 ومن الدليل على غلبة البلغم أن من غلب عليه البلغم كان أبيض اللون لين الشعر رقيقة بليدا بطيئا في الأمور قليل الانتشار والإنعاظ لأن الرطوبة ترخي ذَكَره ويكون كثير الزرع رقيقة مع رطوبة العين والمنخرين وملوحة الفم ورطوبته واسترخاء البدن والتشوق إلى حر الهواء وقلة الكلام ولين ضربان العروق وبرده وأن يرى في منامه الأنهار والبحار والأمطار وما شابه ذلك.

§78 ومن الدليل على غلبة السوداء وهيجانها خضرة اللون وسواده وسواد الشعر وكثرة الحزن والأمراض وطول الفِكَر وبرد البدن ويبس العين والمنخرين وحموضة الفم ويبسه وصغر نبْض العروق وأن يرى في منامه الظلمة والأقدار والتهاويل مَعَما يعتريه من فتور شهوة الجماع وقلة الزرع.

§79 واستدلوا على هيجان هذه المزاجات وفسادها بألوان البول فقالوا: إن من البول ما هو من مائية الدم وهو رقيق ومنه ما يتولد من فضلة الهضم الذي يكون في العروق وسائر الأعضاء وهو أغلظ من الأول، ولون البول شبيه بلون الماء وإنما ينصبغ ويتغير لونه باختلاط الدم والصفراء والسوداء فالأبيض منه يدل على ضعف الحرارة الغريزية في البدن وما اصفرّ قليلا فإنه يدل على أن في البدن حرارة ضعيفة قد غيرت من لونه كما يحدث في لون مَن عدا عدْوا قليلا أو جاع ساعة من النهار فأما ما احمرّ منه فإنه يدل على غلبة الحرارة وقوتها وذلك كما يعرض لمَن عدا عدْوا كثيرا أو صام أياما كثيرة وإن كانت الحمرة إلى الشقرة دلت على نارية وحرارة أقوى من الأولى وإن كان البول إلى لون السواد دل أكثر ذلك على احتراق الدم وبرد القوة ودُنُوّ الأجل وكذلك البول المائي الأبيض ولا سيما في الشباب.

A predominance of phlegm is indicated by (the fact that) the person has a whitish complexion, and soft, sparse hair; he is stupid, and lazy in (managing) his affairs; he is rarely aroused or erect—because moisture slackens his penis—, but he has plenty of thin semen; moreover, his eyes and nostrils are moist, he has a salty, soggy taste in his mouth, and a flabby body; he craves warm air, and talks little; his arteries pulsate calmly and (feel) cold (to the touch); in his sleep he sees rivers, lakes, rain(falls), and the like. § 77

A predominance and agitation of black bile is indicated by (the fact that the person has) a dark, greenish complexion, and black hair; he is often sad or ill, and lost in thought; his body is cold, his eyes and nostrils are dry, he has a sour, stale taste in his mouth, and his arteries pulsate lowly; in his sleep he sees darkness, fateful events, and terrifying objects; moreover, he is affected with a feeble sexual desire, and a lack of semen. § 78

(Physicians) draw conclusions about the agitation and corruption of these mixtures from the colours of the urine. They say: urine that (originates) from the watery (constituents) of blood is thin; urine that originates from the waste products of digestion, which latter takes place in the blood vessels and other organs, is thicker than the first. The (natural) colour of urine resembles the colour of water, yet it is tinted and altered through its blending with blood, yellow bile and black bile. Thus, white urine suggests a weakness of the heat innate to the body; yellow-tinged (urine) suggests a deficiency of warmth in the body as the (likely) cause of that discolouration—this may happen to someone who has been running a little or gone hungry for a few hours of the day; regarding red-tinted urine, it suggests a strong predominance of heat—this may happen to someone who has been running a lot or fasting for several days; if that redness (shades) into a yellowish brown, it suggests fieriness and an (even) stronger (presence of) heat; if the colour of the urine is blackish, it normally suggests that the blood is being burned, the strength is being frozen, and that the end is nigh; the same (final outcome is suggested by) white, watery urine, especially (when it occurs) in youth.[69] § 79

69 In the *Paradise of Wisdom*, the section on uroscopy, drawn from certain "scholarly writings" (*kutub al-ʿulamāʾ*), is considerably longer, see ṬabFir 347,19–355,18. Whilst the short paragraph at hand is almost certainly extracted from that section of the *Paradise*, the likely sources for the latter's expositions on the subject are Hippocrates' *Προγνωστικόν* (UllMed 29), pseudo-Galen's *Περὶ οὔρων* (UllMed 44) and Job of Edessa's (d. c. 832 CE) *Kṯāḇā dTaṗšūrṯā* (RhaCB 42 f., with long [Arabic] fragments ibid. 276–317).

قول الهند في تدبير الصحة

§80 قد بيّنّا ما قالت الروم في تدبير الصحة، فأما الهند فإنها قالت: إن من أراد الصحة فينبغي أن يقوم من نومه في السُبْع الأخير من الليل وليسبغ وُضُوءَه ويلبس أنظف ثيابه ويبدأ بحمد الله والتضرع إليه في حاجاته.

§81 ويستاك من أشجار مُرّة أو حِرّيفة ويكون المسواك رطبا مستويا قليل العُقَد في غلظ الخنصر وطول الشبر وأن لا يكون من شجر مجهولة لأنه لا يؤمن أن يكون سما ولا يجعله عفنا ولا عتيقا ويستاك عَرْضا على الأسنان والحنك واللسان ويغسل الفم في أيام الصيف بماء بارد وفي أيام الشتاء بماء حار، ومن منافع السواك أنه ينقي الفم ويذيب البلغم ويطلق اللسان ويجلوه ويصفي الكلام ويشهي الطعام، ولا ينبغي أن يستاك المتّخِم ولا صاحب القيء ومن به السعال واللقوة والعطش والرمد والخفقان.

§82 فإذا فرغ من سواكه اكتحل بالإثمد، ومن منافعه أنه يصفي الحدقة ويدسمها ويفرح القلب ويجلو القذى، ويكتحل في كل جمعة بالحضض مرةً ليحلب ما فيها من غلظ الرطوبات لأن العين من جنس النار وضد النار الماء، والذي لا ينبغي له أن يكتحل الشبعان ومن قد تقيّأ أو به ورم.

---

70 Ṭabarī drew the core of this chapter (§§80–96) mainly from the sections on 'daily regimen' (*dinacaryā*), 'prophylaxis' (*rogānutpādanīya*) and 'knowledge of liquid substances' (*dravadravya vijñānīya*) in Vāgbhaṭa's (fl. c. 600 CE) *Aṣṭāṅgahṛdayasaṃhitā* (cf. also ṬabFir 565,3), shortening or expanding the material as he saw fit, but largely following the latter's sequential arrangement; besides, Ṭabarī consulted the sections on 'quantitative dietetics' (*mātrāśitīya*), among others, in Caraka's (fl. c. 50 CE) *Saṃhitā*—parallel passages in both Ayurvedic texts are referenced below at the end of each paragraph. In the Indian tradition, generally speaking, day-to-day hygienic duties are considered a matter of medicine as well as religion, and therefore Ayurvedic instructions on that subject often correspond to statements made in the Smṛtis and Gṛhyasūtras, cf. JoIM 45–48.

71 Cf. §§10–19.

72 *wuḍū'* "ritual ablution before prayer" is used here to translate Sanskrit *śaucavidhi* "rite of ablution".

73 This typically Islamic invocation is an obvious addendum.

74 VāgAṣṭ 1/22 no. 1a–b: "A healthy person who values life should get up at dawn (*brāhmamuhūrta*), contemplate the condition of his body, and then perform his ablution". CaSaṃ 1/126 no. 95a: "Wearing tidy garments enhances one's physical charm and reputation […]".

75 VāgAṣṭ 1/22 f. nos. 2a–4b: "Next, he should clean his teeth with a twig (obtained) from

The teachings of the Indians about health regimen[70]

We already explained what the Greeks have to say about health regimen.[71] As regards the Indians, they say (the following): he who wants to stay healthy should rise from his sleep in the final seventh of the night, carefully perform his ablution,[72] put on his tidiest garment, and begin by praising God and imploring His help[73] in times of need.[74] § 80

(Then) he cleans his teeth (with a twig obtained) from bitter or pungent (tasting) trees—this tooth-stick should be fresh, straight, without many knots, as thick as a little finger, and as long as the span of a hand; it must not be (taken) from an unknown tree, for fear of it being poisonous, and not used if it is rotten or old. Teeth, palate and tongue are cleaned widthwise; (afterwards) the mouth is rinsed with cold water in the days of summer, and with warm water in the days of winter. The benefits of the tooth-stick are that it cleanses the mouth, dissolves phlegm, frees and polishes the tongue, clears the speech, and whets an appetite for food; yet it should not be used by someone who suffers from indigestion, vomiting, a cough, facial paralysis, thirst, an eye inflammation, or palpitations.[75] § 81

When he has finished with the tooth-stick, he applies to his eyelids (crushed) antimony, whose benefits are that it clears and lubricates the pupil, gladdens the heart, and rids the eye of impurities. Once a week, so as to make viscid fluids ooze from his eyes, he applies lycium sap—for the eye is akin to fire, and (like) the latter opposed to water. Yet someone who is replete or has just vomited or suffers from a tumour should not use (any such) collyrium.[76] § 82

---

the giant milkweed, the banyan, the black catechu, the Indian beech, the arjun tree and the like, (all of) which are astringent, pungent and bitter in taste; this (twig) should be straight, as thick as the tip of a little finger, and as long as the breadth of twelve fingers. Its top should be (chewed to form) a soft brush before it is used to clean the teeth, without hurting the gums. Those who suffer from indigestion, vomiting, shortage of breath, a cough, fever, facial paralysis, thirst, ulcers of the mouth, and diseases of the heart, eyes, head or ears should avoid using the tooth-stick". CaSaṃ 1/122 nos. 72b–73a: "(Using the tooth-stick) removes a foul smell and a stale taste; it rids tongue, teeth and mouth of dirt, and thereby whets an appetite for food".

76 VāgAṣṭ 1/23 nos. 5a–6a: "Antimony (*sauvīrāñjana*) is good for the eyes, and should be applied daily. The eye, being filled with light (*tejas* [also: 'fire']), is prone to (all kinds of) troubles, especially on the part of phlegm (*śleṣmā*); therefore, in order to drain (the eye), a collyrium made from copper vitriol and barberry (sap) (*rasāñjana*) should be applied once a week". CaSaṃ 1/112 nos. 18b–20a: "Just as objects of tarnished gold can be made to shine by means of oil, cloths and brushes, so the eyes of humans who apply collyria and lubricants become spotlessly bright, like the moon in a clear sky".

§83 ثم ينبغي أن يستعط بأدهان حارة على قدر حاجته إلى ما يزيد في دماغه وما ينقص من فضوله، ومن منافع السعوط أنه يغلظ العنق والعضد ويدسم الوجه ويقوي الحواس ويؤخر الشيب وابيضاض الشعر، ولا ينبغي أن يستعط الممتلئ من الطعام أو الشراب ولا المُتَّخِم ولا من به سعال أو زكام ولا المرأة الحامل.

§84 ثم يتطيب بعد ذلك بما يوافق أيام السنة ويشم الرياحين ويلبس الثياب المطيبة، ومن منافع ذلك أنه يقوي البدن ويفرح القلب ويطيب النَفَس ويذهب البؤس ويزيد في الباه.

§85 ثم يمضغ القرنفل أو جوزبوا أو الكبابة، ومن منافع ذلك أنه يشهي الطعام ويطيب الفم ويذهب بأوجاع الحلق والفم، ولا ينبغي أن يمضغ ذلك من به السل أو مرة هائجة أو خمار.

§86 فإذا فرغ من ذلك كله انتشر في حوائجه وبدأ بلقاء المَشْيَخة والأكابر وأهل الفقه والدين فيقضي حقهم ويستفيد منهم، ثم يقضي حق من يجب عليه حقه من قرابة أو أخ أو سلطان، ثم يتفرغ لطلب معايشه التي بها القوة على ذلك كله.

§87 فإذا كان وقت الغذاء انصرف وعمل البيايام وتفسيره إتعاب البدن بصراع أو عَدْو أو مشى أو ركوب أو غير ذلك من أنواع البيايام، ومن منافع البيايام أنه يكسر الريح وينشطه ويقويه ويخففه ويوقد نار المعدة ويشد المفاصل ويذيب الشحم المفرط والبلغم، ولا ينبغي أن يعمل البيايام الصبي

---

٤ أو : ولا Ox ١٠ و : او Ox ١٢ عدو : غمز (sic) Ox ١٣ ينشطه : سسطه Ox

---

77 VāgAṣṭ 1/23 no. 6b: "Afterwards, the person should make use of nasal drops […]". CaSaṃ 1/120 nos. 57b–63a: "He who uses nasal drops […] protects his hair and beard from turning gray or white; […] the structure of his head and neck becomes more solid, and his face more cheerful and chubby; […] his organs of sense become clear and sharp".

78 CaSaṃ 1/126 no. 96a–b: "Wearing perfumes and garlands stimulates libido, makes the body smell good, enhances longevity and allure, fattens and strengthens the body, pleases the mind, and wards off misfortune".

79 VāgAṣṭ 1/23 nos. 6b–7b: "Afterwards, the person should make use of […] betel leaves. Chewing betel leaves is (however) not suitable for someone who is injured, nor (should it be done by someone) who suffers from bleeding diseases, dryness and redness of the eyes, poisoning, fainting, intoxication, or consumption".

Then he should snuff as much warm oil as is required to profit his brain and to lessen residual matters (in the nasal recesses). The benefits of snuffing are that it fleshes out the neck and the upper arms, fattens the face, heightens the senses, and delays the hair from turning gray or white; yet it should not be done by someone who is filled with food or drink, suffering from indigestion, a cough, or a catarrh, nor indeed by a pregnant woman.[77] §83

Thereafter he perfumes himself with something that is appropriate for the time of year; smells aromatic plants; and puts on scented clothes. The benefits of this are that it strengthens the body, gladdens the heart, sweetens the breath, dispels gloom, and enhances (the desire for) coitus.[78] §84

Then he chews cloves, nutmegs, or cubebs. The benefits of this are that it whets an appetite for food, mends the oral cavity, and abolishes discomfort in the throat and mouth; yet it should not be done by someone who has consumption, an agitated gallbladder, or a hangover.[79] §85

When he has completed all this, he goes out to (fulfill) his daily duties: he first meets the chiefs, elders, and scholars of law and religion,[80] pays his respect, and learns something from them; he dutifully greets neighbours, friends, and government officials; then he sets upon earning his sustenance, which gives him the strength for everything else.[81] §86

When mealtime has come, he returns (home) and practises *bayāyām*, which can be translated as 'exerting the body'[82] through wrestling, running,[83] walking, riding, or another such kind of physical exercise. The benefits of *bayāyām* are that it breaks up flatulence, enlivens, strengthens and lightens (the body), ignites the fire of the stomach, supports the joints, and melts excess fat and phlegm; yet it should not be practised by a child or an old §87

80 *ahl al-fiqh wad-dīn* "scholars of law and religion" involves typically Islamic terms and concepts; the Sanskrit here has *vipra* "members of the Brahmin caste".

81 VāgAṣṭ 1/26 f. nos. 20a–24a: "All (human) activities are meant to further the happiness of all living beings; such happiness is based on proper moral conduct, which should therefore be observed always and by everyone. Friends are to be met with affection and kindness; [...] refusing to learn from the sages (*dṛgviparyayā*) is sinful; [...] gods, cows, Brahmins, the elders, the physician, the king, and guests must be treated with utmost respect".

82 *bayāyām*, explained by Ṭabarī with 'exerting the body' (*itʿāb al-badan*), is a straightforward transliteration of Sanskrit *vyāyāma* "exertion, gymnastic exercise".

83 *ʿadw* "running" is an emendation of the strange manuscript reading غمز.

ولا الهرم ولا من به تخمة، ومقدار البيايام القصد فإن الإكثار منه يورث المرة والعطش والدوار والسل والسهر والسعال.

§88 فإذا فرغ منه مرخ بدنه بعد ذلك بأدهان مطبوخة على قدر حاجة كل إنسان وفي كل زمان، ومن منافع ذلك أنه يبطئ بالهرم ويذهب بالإعياء ويديم الصحة ويلين الجلد ويطيب النوم، وأفضل التمريخ أن يكون في الرأس والقدمين والأذن فما كان منه في الرأس فإنه يسود الشعر ويقوي الحواس وما كان منه في الأذن فإنه يدفع وجع الأذن والعنق والوجه وما كان منه في القدم طيّبَ النوم وأذهب الشحوبة والإعياء وزاد في الباه، ولا ينبغي أن يتمرخ صاحب البلغم والتخمة ولا بعد شرب دواء الإسهال والقيء.

§89 فإذا فرغ من التمرخ دلك جسده بالنخالة دلكا جيدا وغسله في الصيف بالماء البارد وفي الشتاء بالماء الحار، ومن منافع الاغتسال أنه يوقد نار المعدة ويزيد في الباه ويديم الصحة وينقي البدن ويفرح القلب، ولا ينبغي أن يغتسل من به رمد أو لقوة أو نفخة أو زكام أو

---

٦ منه[2] : – Ox   ١١ ينقي : يسى Ox

---

84 *mirra* "bile" here replaces Sanskrit *pitta*.

85 VāgAṣṭ 1/24 f. nos. 10a–13b: "Physical exercise makes (the body) light, enables to do hard work, kindles the digestion, melts (excess) fat, and firms and stabilizes the physique; it should (however) be avoided by victims of bile- or wind-related diseases, by those who suffer from indigestion, and by children and old people. Those who are strong and (regularly) consume fatty foods, as well as (those who find themselves) in cold temperatures or in the season of spring, should exercise only to half of their capacity; others, and in different seasons, should practise it moderately. […] Thirst, emaciation, severe shortage of breath, bleeding diseases, exhaustion, weariness, coughing, fever, and vomiting are caused by excessive exercise". CaSaṃ 1/152 nos. 32a–33b: "Physical exercise brings lightness, the ability to work, resistance to discomfort, an alleviation of disordered humours (*doṣa*), and it stimulates the digestive power; excessively practised, it leads to exertion, exhaustion, consumption, thirst, sudden bleeding from the mouth or nose, asthma (*pratāmaka*), coughing, fever, and vomiting".

86 *balġam* "phlegm" here replaces Sanskrit *kapha*.

man, nor by someone who suffers from indigestion. The (right) measure of *bayāyām* (lies in) moderation—too much of it causes (an increase of) bile,[84] thirst, dizziness, consumption, insomnia, and coughing.[85]

Then, when he has accomplished that, he embrocates his body with as much cooked oil as is warranted by his individual needs and the particular time (of year). The benefits of this are that it slows down (the process of) ageing, dispels weariness, perpetuates health, softens the skin, and improves the sleep. Embrocation is most propitiously applied to the head, the feet and the ears—applied to the head, it blackens the hair and heightens the senses; applied to the ears, it shelters them, as well as the neck and the face, from pain; and applied to the feet, it betters the sleep, removes pallor and fatigue, and enhances (the desire for) coitus. Yet it should not be used by someone who is dominated by phlegm[86] or (suffering from) indigestion, and not after having taken a purgative or vomitive drug.[87] § 88

When he has finished with oiling himself, he thoroughly rubs his body with bran, (then) washes with cold water in summer or warm water in winter. The benefits of washing are that it ignites the fire of the stomach, enhances (the desire for) coitus, perpetuates health, cleans the body, and gladdens the heart; yet it should not be done by someone who suffers from an eye inflammation, facial paralysis, bloating, a catarrh, diarrhoea, or hallucinations, nor § 89

87 VāgAṣṭ 1/24 nos. 8a–9b: "Embrocation (*abhyaṅga*) should be done daily. It wards off old age, fatigue and wind (diseases), improves the sight, nourishes the body, (grants) long life and good sleep, and makes the skin smooth and firm. It should be applied especially to the head, the ears and the feet. It must be avoided by those who suffer from an aggravation of phlegm or indigestion, as well as by someone who has just undergone purification therapies". CaSaṃ 1/124 f. nos. 81a–90b: "He who regularly applies oil to his head will not suffer from headache, (nor be affected by) baldness or graying of the hair; [...] (rather) his hair becomes black, long and deeply rooted; his organs of sense will work properly; [...] his sleep will be sound. [...] If oil is dropped regularly into the ears, it protects from wind-induced corruptions, wryneck, lockjaw, hardness of hearing, and deafness. [...] By oiling the feet, roughness, immobility, dryness, fatigue, and numbness are cured instantly".

اختلاف أو هذيان ولا يغتسل أحد إلا وعليه مئزر يواري عورته، ثم يتطيب بما يوافق بدنه وزمانه من الطيب.

§90 ثم ينظر فإن اشتهى الأكل وخلا جوفه وصفا جشاؤه أَكَلَ وإلا أَخَذَ ما يُمْرِئ ويقوي وينقي معدته أولا، وينبغي له أن لا يجامع ولا يتجشأ ولا يعطس إلا ووجهه من تلقاء صدره، ويبدأ قبل الأكل فيطعم أَبَوَيْه وقراباته وأهل الحُرْمة والحاجة ثم يطعم البهائم والطير ثم يأكل، ولا يطعم من طعام متغير ولا محترق ولا من طعام لا يعرفه أو يتقزز منه ولا من طعام لا يثق به ولا من طعام الصدقة، ولا يأكل في ظلمة ولا في شمس ولا مُمْسِيا ولا تحت شجرة مجهولة ولا على الفراش ولا من طعام قد أكل قبله ما يضاده ولا يأكل مستعجلا ولا بطيئا.

§91 وليحذر أن يمنع أو يردع ما ينبعث أو يهتاج من بدنه مثل العطاس والجشاء وشهوة الطعام والشراب والنوم والباه وما ينبعث من البول والبراز والعرق والقيء والبزاق والريح وما أشبه ذلك، لأن حبس الريح يورث الحُصْر والأُسْر وظلمة العين ووجع الفؤاد، وحبس البول يورث

---

٧ في[2]: – Ox

---

88 VāgAṣṭ 1/25 f. nos. 15a–18b: "Rubbing the body with fragrant powders (*ud-vartana*) mitigates phlegm, liquefies fat, tightens the limbs, and embellishes the skin. Washing with water (*snāna*) improves the appetite and the desire for coitus, (extends) the lifespan, (gives) energy and strength, removes itching, dirt, exhaustion, sweat, stupor, thirst, burning sensations, and sin (*pāpmā*); [...] washing should not be done by those who suffer from facial paralysis, diseases of the eyes, mouth and ears, diarrhoea, flatulence, catarrh (*pī-nasa*), and indigestion, nor by someone who has just eaten". CaSaṃ 1/126 no. 94a–b: "Washing (*pavitra*) purifies, stimulates libido, gives life, removes fatigue, sweat and dirt, and brings strength and vitality to the body"; ibid. 1/173 no. 19: "One should not wash [...] when naked".

89 *ahl al-ḥurma* is ambiguous in Arabic: it may be interpreted as 'holy people' or as 'forbidden people'. In the former case, which I consider much more likely in the given context, the reference would be to "priests" (*brāhmaṇa* or *vipra*); in the latter case, it might refer to "untouchables" (*dalita*). The Ayurvedic parallel texts are not helpful here.

90 *ṭaʿām aṣ-ṣadaqa*, lit. "charity food", involves the Islamic concept of almsgiving (*ṣadaqa*); the Sanskrit here reads "(food distributed in an) asylum or sacrificial ceremony" (*sattragāra*).

91 The extant Ayurvedic sources are much more elusive here. See, however, VāgAṣṭ 1/26 no. 19a: "One should take food only after (the previous meal is) digested, and (eat only) what is suitable, in limited quantities"; ibid. 1/29 no. 35b: "One should not sneeze, laugh or yawn without covering the mouth"; ibid. 1/30 nos. 41a–43b: "One should not sneeze,

must anybody wash without wearing a waist-wrapper to hide his genitals. Then he perfumes himself with a fragrance that is appropriate for his body and the time (of year).[88]

Then he follows this: if he feels an appetite for food, if his belly is empty and his belching pure, he eats; if not, he first reaches for something that digests easily, and that strengthens and cleanses the stomach. He must not (eat) during sexual intercourse, nor burp or sneeze unless he faces chestwards. Before he has his own meal, he feeds his parents and his family, the priests[89] and the paupers, the cattle and the birds—(only) then he eats himself. He touches no food that is spoiled or burned, none that he does not recognize or trust, none that he finds repulsive, and none that has been donated;[90] he does not eat in the dark, not in the sun, not late in the evening, not under an unknown tree, and not in bed; he (also) refrains from anything that is incompatible with what he just had; and he eats neither in a hurry nor at a dragging pace.[91] §90

One must be careful not to repress or retain anything that comes forth from the body or irritates it, such as an impulse to sneeze, belch, take food and drink, sleep, or have sex; the same goes for (an urge) to urinate, defecate, perspirate, vomit, expectorate, break wind, and the like. (This is so) because withholding wind leaves constipation, ischuria, darkening of the §91

---

burp, cough, sleep, eat or engage in sexual activities without assuming a decent posture, [...] avoid food that has been given by enemies, or distributed in an asylum or sacrificial ceremony, or handed out by prostitutes or merchants"; ibid. 1/130 f. nos. 35a–38b: "Food, to be consumed at the proper time, should be familiar (to a person), unsoiled, wholesome, [...] and easy to digest; he should eat it with due attention, [...] neither too quickly nor too slowly, and (only) after having fed parents, gods, guests, children, teachers and all others present in the house; [...] he should eat (only) what he likes, in the company of people he knows, and what has been given to him by (servants) who are cleanly and trustworthy"; ibid. 1/134 no. 55a: "The ideal time for eating is when [...] belchings are pure and hunger manifest". CaSaṃ 1/175 no. 20: "One should not take food [...] without offering some to the gods, the departed ancestors, the teachers, the guests, and the dependants, [...] nor (should one eat) when surrounded by untrustworthy, filthy or starving people, nor (when that food is served) on unclean plates or in suspicious locations, nor (when the food itself) is dirty or offered by opponents". Ṭabarī's reference to putting food aside for "the paupers, the cattle and the birds" is echoed also by al-Bīrūnī (d. 440/1048), in his work on India, as part of the 'universal duties' of a Brahmin, see BeInd 2/134.

هذه كلها ويورث الحصاة ووجع المثانة، وحبس البراز والنجو يورث هذه كلها لأنها إذا حُبِسَت من أسفل طلبت المخرج من فوق، وحبس الجشاء يورث الفواق، ومنع النوم يورث الكسل والثقل في الرأس والعين، وحبس الباه يورث وجع المثانة والفؤاد ويورث الحصى وسيلان الزرع.

§92 وقالت الهند: إن الإكثار من الأشياء الحارة يذهب بالقوة وييبّس البطن ويغير اللون، والإكثار من الأشياء الدسمة يورث الكسل والرطوبات ويذهب بالشهوة، والإكثار من الأشياء المالحة يضر بالبصر ويهزل البدن، والإكثار من الحامض يجلب الهرم.

§93 وقالت: إن علل هيجان البلغم الإكثار من الحلو والدسومات وكثرة النوم والدعة والخبز المعجون بالسمن واللبن ولحوم نبات الماء، وعلل هيجان المرة الإكثار من الأطعمة الحِرّيفية الحامضة وكثرة شرب البلاذر والأدوية الحارة وكثرة الجماع والغضب والخوف الشديد والحسد.

---

٥ يبس : ييبس Ox ٧ يجلب الهرم ṬabFir 568,3 : – Ox

92 VāgAṣṭ 1/45 no. 1a–b: "The urge to break wind, defecate, urinate, sneeze, drink, eat, sleep, cough, breathe on exertion, yawn, vomit, and ejaculate semen must not be suppressed"; ibid. 1/45 no. 2a–b: "Suppressing the urge to break wind causes intestinal tumours, upward movements inside the alimentary tract, (abdominal) pain, exhaustion, constipation, loss of vision, deterioration of the digestive power, and heart diseases"; ibid. 1/45 nos. 3a–4a: "Suppressing the urge to defecate causes pain in the calves, a running nose, headache, upward movements of air, cutting pain in the rectum, an oppressive sensation in the region of the heart, vomiting up feces, as well as the problems mentioned earlier"; ibid. 1/46 nos. 4b–5a: "Suppressing the urge to urinate causes cutting pain all over the body, urinary stones, severe pains in the bladder, penis and groins, as well as the problems mentioned earlier"; ibid. 1/46 nos. 7b–8b: "Suppressing the urge to belch causes loss of appetite, tremors, tightness around the heart and in the chest, flatulence, coughing, and hiccups"; ibid. 1/47 nos. 12b–13a: "Suppressing the urge to sleep causes delusions, a heavy sensation in the head and in the eyes, lassitude, excessive yawning, and squeezing pains all over the body"; ibid. 1/49 nos. 19b–21a: "Suppressing the urge to ejaculate semen causes it to leak, (further) painful and swollen genitals, fever, discomfort in the region of the heart, ischuria, cutting pain in the body, inguinal and scrotal hernia, urinary stones, and impotence". Very similar expositions are found in CaSaṃ 1/146–150 nos. 3a–4b, 6a–b, 8a–b, 10a–b, 12a–b, 18a and 23a.

eyes, and central chest pain; withholding urine causes all these (problems), and (moreover) stones and pain in the bladder; withholding feces and excrement causes all these (problems), too—for if (the path of) fecal matter is blocked from below, it seeks an exit from above; withholding belches leaves hiccups; suppressing sleep leaves sluggishness, a heavy head and heavy eyes; and withholding from coitus leaves pain in the bladder and central chest, (vesical) stones, and semen leakage.[92]

The Indians (further) say: overmuch (consumption) of hot stuff removes strength, desiccates the belly, and pales the complexion; overmuch fatty stuff causes sluggishness, (generates) fluids, and spoils the appetite; overmuch salty stuff damages the sight and emaciates the body; and overmuch sour (stuff) accelerates (the process of) ageing.[93] § 92

They say: phlegm may be agitated for the (following) reasons—excessive (consumption) of sweet or fatty stuff; (too) much sleep; idleness; (indulging in) bread that has been kneaded with clarified butter and milk; or (eating) the flesh of aquatic plants. Bile may be agitated for the (following) reasons—overmuch pungent (or) sour food; frequent ingestion of marking-nuts or (other) hot drugs; immoderate sexual intercourse; anger; intense fear; or envy.[94] § 93

93 Ayurveda distinguishes six basic 'tastes' (*rasa*): sweet, sour, salty, bitter, pungent, and astringent, see e.g. VāgAṣṭ 1/143 no. 1a–b and CaSaṃ 1/46 no. 65a–b. Ṭabarī's 'hot' and 'fatty' substances do not fit into this category, even though these qualities are of course recognized by Ayurvedic authors, either as *doṣa* attributes (hot) or simply as organic constituents (fatty). Overall, there is no obvious link between the statements Ṭabarī makes in this paragraph and the scattered remarks on the subject in his Ayurvedic sources. However, as regards the negative effects of 'salty' and 'sour' substances, some vague parallels can be found in the *Aṣṭāṅgahṛdayasaṃhitā*, viz. "Generally, salts are bad for the eyes" (VāgAṣṭ 1/149 no. 34b), "Salt (*lavaṇa*), used in excess, [...] withers the body" (ibid. 1/145 no. 13b); further, "Sour stuff (*amla*), used in excess, [...] fatigues the body and diminishes its strength" (ibid. 1/145 no. 11b).

94 Here again, as in § 92, it is difficult to establish a clear-cut textual link between Ṭabarī's expositions and his Ayurvedic sources, which are overwhelmingly concerned with *effects* (rather than causes) of *doṣa* aggravation. Only one more or less direct literary correspondence to Ṭabarī's statements in this paragraph can be found in the *Carakasaṃhitā*, viz. "All foods of sour taste aggravate bile (*pitta*); [...] all foods of sweet taste aggravate phlegm (*śleṣmā*)" (CaSaṃ 1/491 no. 4). It seems reasonable, in both cases (§§ 92 and 93), to allow moreover for the possibility of *oral* transmissions.

§94 وقالوا: إن الماء حياة لكل ذي روح ونبت، وإن من المطر ضربيْن: ضرب نهري وضرب بحري، لأنه ربما لطف ماء البحر فصعد إلى الهواء ثم صار مطرا، ويُستدل على النهري بأن يوضع قِدْر من فضة فيها أرز مطبوخ ويكون ذلك في المطر فإن لم يتغير رائحته فهو الماء المبارك السمائي النهري وإن تغير فهو بحري، وإذا كان نهريا أُخِذَ في كرابيس جدد وصير في خزف أو زجاج ويُعْهَد بالترويق، ويجنَّب ما كان متغيرا أو مستنقعا في الأرض فإنه يستفيد من قوة تلك الأرض، وإنما يُعْرَف خفة الماء وثقله من البلدان فإذا كانت الأرض قاعًا حَرَّةً قليلة العفونة فإن ماءها خفيف وما كان من أرض شجراء كثيرة العفونة فإنه ثقيل، ويُجنب الماء الذي فيه طحلب وديدان وحيات، وقالوا: إن شرب الماء البارد قبل الطعام يهزل البدن ويطفئ نار المعدة وشربه بعد الطعام يسخن البدن ويزيد البلغم.

§95 قالوا: فأما الشراب فمبارك في الدنيا والآخرة، وعجبا لمن كان شرابه شراب الكرم وأكله الخبز واللحم ثم اقتصد في أكله وشربه وجماعه ولعبه كيف يمرض وكيف يموت، فإن الشراب يفرح أوج القلب ويجترئ به الجبان ويسخو به اللئيم ويحسن عليه اللون ويشهي الطعام، وأما الإكثار منه فإنه مفتاح الفتن والآثام والفضائح ويميت مع هذا الشهوات ويغير العقل الذي به فرق ما بين الناس والبهائم، ولا ينبغي أن يشرب منه الملك ولا وزيره ولا الطبيب الذي به يؤتمن على أنفس الناس ولا خادم الملك ولا قاضيه الذي ينفذ حكمَه في المُهَج والأموال، وينبغي لمن أكثر منه أن يغتسل في ماء حار ويستقبل جِرْية الماء ويجلس في مَظالّ معمولة من الخلاف

٥ يجنب : تجنب Ox ٦ حرة : جرا Ox

95 VāgAṣṭ 1/53 ff. nos. 1a–9a: "Rain water (*gaṅgāmbu*), fallen from the sky and come into contact with sunlight, moonlight and wind, enlivens, satiates, comforts the heart, refreshes, and stimulates the mind; it is thin, almost tasteless, slightly sweet, cooling, easily digested, and of nectar-like quality; […] (particularly) good to drink is the kind that does neither sodden nor discolour cooked rice placed in a silver pot. Other kinds (of water) relate to the sea (*sāmudra*) and should not be used for drinking except in autumn (*āśvayuja*). Rain water that has been collected in a clean jar and that does not change (its taste or colour) can always be drunk. Otherwise, (one can drink) water from the earth, (for) this is just like rain water, (provided that) it has been fetched from places that are clean, open, exposed to sunlight and wind, and whose soil is black or white. Water that is dirty, or mixed with slush, algae, weeds and leaves, or hidden from sunlight and wind, or thick, or heavy, or frothy, or worm-infested, or hot (in essence) but cold to the teeth, or gathered from unseasonal or premature rainfalls, or contaminated with the webs, spittle and excretions of spiders and the like, or corrupted by other

They say: water (means) life for all soul-possessing creatures as well as for plants. There are two kinds of rain—one is fluvial, the other maritime; for sea water is sometimes (so) thin that it rises up into the air and later becomes rain. One can identify the fluvial (kind) by putting a silver pot (filled) with cooked rice into the rain—if the smell of that (water) does not change, then it is the blessed, heavenly, fluvial (kind); if (the smell) changes, then it is the maritime (kind). Fluvial (rain water) is caught in new cotton cloths, (then) poured into a jar made of clay or glass, (then) thoroughly filtered. One must keep away from (water) that is dirty or that stagnates on the ground, because (such water) adopts any (noxious) powers of the soil. One can tell light water from heavy water by (looking at the character of) the land—if it is flat, stony and almost free of putrefying matter, then its water is light; if it is bosky and full of putrefying matter, then (its water) is heavy. Avoid water in which there are algae, worms or snakes. They say: drinking cold water before a meal emaciates the body and extinguishes the fire of the stomach; drinking it after a meal heats the body up and increases phlegm.[95] § 94

They say: as for wine, it is blessed in this world and in the hereafter.[96] How strange is it that those who are accustomed to (a diet of) grape wine, bread and meat, then (suddenly) embrace austerity with regard to food, drink, sex and partying, become ill and die! Wine (in moderation) surrounds the most lofty heights of the heart with joy, it lends courage to the coward and generosity to the miser, it embellishes the complexion and whets an appetite for food; (consumed) excessively, it opens the door to temptations, sinful behaviour and scandalous acts, and on top of that deadens (salutary) desires and deranges the mind, which after all distinguishes humans from beasts. Those who should abstain from wine are the king, his minister, the physician to whom the souls of people are entrusted, the king's servant, and the king's judge whose rulings impact on lives and goods. He who had too much wine should wash himself with warm water or stand in a stream; (then) sit in a hut made of willow or poplar (branches) and myrtle, apple or pomegranate § 95

noxious matters, is unsuitable for drinking. Water from clean rivers (*nadījala*) that flow swiftly into the Arabian Sea (*paścimā*) is good to drink"; ibid. 1/56 no. 15a: "People who drink water during, or after, or before a meal, will remain unchanged, or become stout, or emaciate, respectively". Cf. also the relevant section in SuSaṃ 1/323–333 nos. 1–46b.

96 *mubārak fī d-dunyā wal-āḫira* "blessed in this world and in the hereafter" is a typically Islamic turn of phrase and, insofar as it involves the concept of a (heavenly) afterlife, no doubt an addition. Wine is generally prohibited in Islam (except for medical purposes), but in Paradise "rivers of wine delight the drinkers" (*Qur'ān* 47:15).

والصفصاف والآس وأطراف التفاح والرمان ويكون مجلسه على شط نهر أو بِرْكة ويرش فوق مظلّته ماء ورد ويطلي جسده بكافور وصندل وماء ورد وماء الخلاف والزعفران ويتروح بمراوح مطيبة بهذه الأفاويه ويشم الأشياء الباردة، قالوا: ومن أراد أن يدعها فلا ينبغي أن يقطعها في يوم واحد بل يقلل منها أولا أولا ثم يشرب في كل يوم قدحا ثم يدع ذلك ويشرب نبيذ الزبيب والسكر وغيرهما ويأكل خبزا مثرودا في الشراب أياما ثم يدعها.

§96 وقالوا: إن الأطعمة والأشربة فيها حياة الحيوان وبها تقوى نار المعدة، فإن قلت الأطعمة والأشربة فيها انطفأت نارها وكذلك إن كثرت عليها مثل النار التي إن وضع عليها حطب كثير جزل رطب أطفأها وإن وضع عليها قليل خف لهبها، وقالوا: وقد يكون طعام فيه شفاء لأهل بلد وهو لغيرهم مرض ويكون طعام هو وحده غذاء فإن أكل مع شيء آخر صار داء وقد صدقوا في ذلك، وإن السمك الطري يؤخذ وحده حارا غذاء فإن أكل مع شيء آخر صار داء، فإن الطري من السمك

---

١ التفاح : الـ●ـفاح Ox ٧ انطفأت : انطفت Ox ١٠ داء + : وقد صدقوا في ذلك Ox

---

97 Suśruta, Caraka and Vāgbhaṭa all have extensive sections on alcoholism (*madātyaya*), see SuSaṃ 3/305–318, CaSaṃ 4/385–432 and VāgAṣṭ 2/286–305; for the sake of clarity, the following Ayurvedic parallel passages are arranged according to the narrative sequence of Ṭabarī's much shorter and disjointed expositions on the subject of wine. Into the opening sentence of the paragraph at hand, Ṭabarī seems to have condensed lengthy panegyrics on the 'divine' origin and ritual significance of wine (CaSaṃ 4/385 nos. 3a–10b, VāgAṣṭ 2/294 f. nos. 54a–67b)—for example, CaSaṃ loc.cit. "The drink which, from the days of yore, is adored by the gods, [...] respected by the priests, and which sustains the Vedic sacrifices", or VāgAṣṭ loc.cit. "That which [scil. wine] was born from the Great Ocean filled with all herbs and churned by the gods and by the demons", and so on. Next, CaSaṃ 1/152 nos. 34a–35b: "A person should not excessively exercise, laugh, talk, travel on foot, engage in sexual intercourse, or wake through the night unless he is habituated to it; if he indulges immoderately in these and other activities, he will perish like the lion that tries to drag away an elephant". Next, CaSaṃ 4/398 no. 62a–b: "Alcohol in moderation invigorates, whets an appetite for food, promotes the digestion, uplifts the heart, and embellishes the voice and the complexion"; ibid. 4/397 nos. 56a–57b: "Alcohol causes illusions, fear, grief, anger, and (possibly) death, as well as conditions like insanity, intoxication, fainting, epilepsy, and convulsions; whenever the mind is impaired, all (kinds of) evil deeds materialize, which is why those who are acquainted with the adverse effects of alcohol strongly condemn its consumption"; VāgAṣṭ 2/295 nos. 60b–61a: "By drinking (wine) as much as they desire, brave men fight with valour (even) an army of shape-changing nymphs (*ap-saras*), and warriors sacrifice their lives just as (easy as) straw (snaps)". Next, SuSaṃ 3/314 f. nos. 55b–62a: "(He who suffers from alcohol-induced fever [*dāha*]) should rub his body

twigs, close to the bank of a river or near a pond, and besprinkle the roof of his shelter with rose water; (further) rub his body with camphor (leaves), sandalwood, rose water, willow water, or saffron; ventilate himself using a fan that has been perfumed with suchlike aromatics; and smell cold stuff. They say: he who wants to give up wine should not quit from one day (to the next) but rather gradually reduce (the amount), until he is down to one cupful a day which he then replaces by fermented raisin juice with sugar, or something like that; (alternatively), he may eat for a few days crumbled bread moistened with wine, then let it go.[97]

They say: food and drink give life to (all) creatures, and strengthen the fire of the stomach. If there is not enough food and drink in the stomach, its fire is extinguished; the same (happens) if there is too much. (This is just) like a fire which is stifled if one puts a lot of thick, damp chunks of wood on it, but whose flame (also) dies down if it gets only little. They say: sometimes (an item of) food holds a cure for the inhabitants of (one) country, whereas for others (it holds) a disease; (certain) food may be wholesome on its own, but become noxious when eaten together with something else. And they are right in this matter, for fresh fish, taken hot and on its own, is wholesome, but becomes noxious when eaten together with (certain) other stuff. The thing §96

with a paste made from sandalwood and water, as cooling as moonlight or hailstones; he should lie in a hut close to cold water, or sleep in a place that is near to a pond; [...] he should walk about on lawns or in gardens, and expose himself to mild currents of air and to fans perfumed with the aroma of water lilies, lotuses or hornwort, waved in front of him by women; he should pour over himself tepid water enriched with sandalwood, lotus flowers and vetiver roots, or play in a pool filled with clean, fresh water and containing lotuses and lilies, after having had his skin massaged by the hands of women; [...] exhausted (by overmuch drinking), the patient should rest in the inner chamber of a house that is equipped with fountains, [...] airy from all sides, whose floors have been sprinkled with flower-scented water, [...] and which has been decorated with garlands of red lotus, swinging in a fanned breeze"; cf. CaSaṃ 4/419 nos. 152a–164a and 427 nos. 191a–193b. Next, CaSaṃ 1/153 nos. 36b–37b: "(To quit an addiction), a person should give up, on the first day, a quarter of the unwholesome substance; on days two and three, another quarter; on days four, five and six, another quarter, such that on day seven, the schedule can be completed"; VāgAṣṭ 2/291 no. 35a–b: "(In alcoholism with agitated phlegm [*śleṣmā*]), the patient should drink [...] fermented sugarcane juice mixed with dry flour (*tarpaṇa*)". Finally, the advice that "the king, his minister, the physician [...], the king's servant, and the king's judge" should abstain from wine, is given nowhere in Ṭabarī's Ayurvedic sources, nor indeed is it consistent with mainstream Hindu or Buddhist ethical prescriptions; cf. for a modified counsel ṬabFir 572,23–573,1.

هذه قصته: هو غذاء يغذو البدن وحده ويزيد في الباه وإن أخذ مع الجبن واللبن صار داء دويا، ومثل الألبان والرواصيل والبليلج فإنها غذاء لأهل الجبال وسم لأهل العراق، فهذه جملة ما قالت الهند في ذلك.

في معرفة قوى الأشياء والاستدلال عليها بمذاقتها

من قول الروم

§97 قال جالينوس: إن كل شيء يتربى به الإنسان فهو غِذاء وكل شيء يغير الطبيعة فهو دِواء، ومن الأدوية ما هو سم للناس وغذاء للطير مثل الفربيون فإنه غذاء للزرازير ومنها ما هو غذاء للناس وسم لبعض الحيوانات مثل دهن الزيت فإنه سم للخنافس.

§98 ولكل شيء مذاقة يُستدل بها على قوته وله خاصية لا تُعْرَف علتها إلا بالتجربة مثل حجر المغناطيس الذي يجذب الحديد، ومن الأشياء ما يقصد للمثانة فيخرج حصاها مثل العقرب المحرقة بالنار وبزر الكرفس الجبلي، ومنها ما يقصد القلب فيقتل مثل السم ومنها ما يتبع السم فيقتله بإذن الله تعالى مثل الجدوار والترياقات، ومنها ما يعلق في العنق فيمنع من وجع اللهاة مثل الحلتيت، ومنها ما يدخن به البيت فيطرد الحيات مثل قرون الأيائل، ومنها ما يعلق في العنق فيشفي من الصرع

٢ سم : سلم Ox ٦ يتربى TabFir 399,14 : يبرا Ox ٧ الفربيون : الفوينون Ox ٩ تعرف علتها : يعرف علما Ox ١١ يتبع : يتبعه Ox ١٢ الجدوار : الجداور Ox || وجع : وجه Ox ١٣ الأيائل : الايايل Ox

98 The term *rawāṣīl* is hard to pin down lexicographically, and even in medico-pharmaceutical texts it is rather elusive. Rhazes (d. 313/925), in one of his dietetic treatises, mentions *rawāṣīl* as part of a chapter-heading that runs as follows: "On vinegar-based pickles, *rawāṣīl*, mature cheese, barley water, pomegranate juice, cauliflower, olives, pickled vegetables, and the like" (*Fī l-kawāmīḫ war-rawāṣīl wal-ǧubn al-ʿatīq waš-šilmāb wan-nārāb wal-qunnabīṭ waz-zaitūn wal-muḫallalāt wa-naḥwihā*), see RhaMan 31,35 f.; in the chapter itself, all these items, except for *rawāṣīl*, are explicitly reiterated, and it therefore seems likely that the term is semantically close to, if not synonymous with, *kawāmīḫ* "vinegar-based pickles".

about fresh fish is this: on its own, it is (an item of) food which nourishes the body and promotes coitus; but taken with cheese or milk, it turns into a nasty disease. Likewise, milks, *rawāṣīl*[98] and beleric myrobalans are nourishment for mountain dwellers, but poison for those who dwell on the shore.[99] And this is all that the Indians have to say about the subject.

On knowing the powers of substances
and on inferring them from their taste
—after the teachings of the Greeks—

Galen said: anything by which man grows is (considered) nutrition; anything that manipulates nature is (considered) medication.[100] Some drugs are poison for humans but nourishment for birds, like (resin) spurge which is eaten by starlings; others are nourishment for humans but poison for certain animals, like olive oil which is toxic for dung beetles. §97

Every substance has a taste from which one can infer its power; it may (also) have a special property whose cause is unknown but (whose effect is proven) by experience, such as the lodestone which attracts iron. Some substances attain the bladder and dislodge its stones, like scorpion ash or mountain celery seeds; some attain the heart and kill it, like poison, whilst others track poison and kill it (in turn)—with the permission of God the Sublime—, like zedoary or theriacs; some are hung around the neck and prevent uvula pain, like asafoetida resin; some are used to fumigate the house and chase away snakes, like staghorn; some are hung around the neck and—according to a §98

99 Ṭabarī's expositions in this paragraph are hard to substantiate in his Ayurvedic sources. The following statements, however, may be considered relevant: "Food sustains the life of (all) living creatures" (CaSaṃ 1/565 no. 349a); "Strength, health, longevity, and vital breathing depend on the power of digestion—when supplied with fuel in the form of food and drink, this power is encouraged, but it dwindles when deprived of it" (ibid. 1/563 no. 342a–b); "Fish is [...] nourishing [...] and enhances sexual desire" (ibid. 1/508 no. 82a–b). On the concept of a fire-like agency (*agni*), which in Ayurveda is subdivided into five kinds and which is vital for a proper digestion, see VāgAṣṭ 1/406 no. 59a–b and 410 nos. 70b–72a.

100 GalKü 11/380 (*Περὶ κράσεως καὶ δυνάμεως τῶν ἁπλῶν φαρμάκων*, cf. FiCG no. 78): "Medicamentum sane omne id dicimus, quod naturam nostram alterat; sicut, puto, et nutrimentum, quicquid substantiam auget".

على ما ذكر جالينوس وهو دواء يقال له فاونيا وهو في قدر باقلاة، ومنها ما يعلق في الفخذ فينفع من وجع القولنج مثل زبل الذئب.

§99 وقد قال أرسطاطيليس: إن المذاقات ثمانية أولها الحلاوة ثم المرارة ثم الملوحة والحموضة والحرافة والعفوصة والبشاعة والزِفِرة، ولهذه المذاقات أفاعيل عجيبة ظاهرة في الجسم أنا ذاكرها.

§100 فأقربها إلى الطبيعة المعتدلة الحلاوة لأنها تتركب من جزء من حرارة تمتزج بجزء من رطوبة فإن زادت إحدى القوتيْن أو نقصت تغير الطعم على قدر ذلك، فأما المرارة فإنها تحدَث من حرارة ويبس، وكل مالح فهو حار يابس أرضي وهو دون المر ودون الحِرِّيف في الحرارة، والملوحة مثل البحر الذي تنشف الشمس ما لطف منه ويبقى ما غلظ وخاصية الملوحة وفعلها أن تغوص في البدن وتذيب الرطوبات وتحفظ الاعتدال، وكل قبّاض بارد يابس مثل الثمار فإنها صلبة عفصة أولا مثل العنب والتفاح والرمان فإذا استحكمت بحرارة الشمس وطلوع القمر عليها اعتدلت واحْلَوْلت، وكل ما كان حلوا محضا فهو غاذٍ وكل ما كان بين الحلو والمر فهو يغذو غذاء يسيرا، وكل حامض فإنه بارد لطيف ينقي مجاري البدن ويجلو، وكل عفص فإنه أرضي ييبّس، وكل حِرِّيف فإنه ناري حار يلطف الفضول، وكل دسم فهو حار رطب يلين البدن ويرخيه من غير تسخين، وكل شيء يزيد في الزرع فهو حار رطب أو حار نفّاخ وكل ما قطع الزرع فإنه يقطعه بحرارة ويبوسة مثل السذاب والخردل والشهدانج والفنجنكشت وإما بالبرد واليبس مثل بزر الخشخاش الأسود وبزر الخس.

---

١ فاونيا: فاوِنيا Ox ٣ أرسطاطيليس: ارطاطيليس Ox || ثمانية: يمنيه Ox ٤ العفوصة: العنوصة Ox ||
الزفرة: الدسومة Ox ṬabFir 358,7 ٥ من جزء من: في جزو Ox ٦ إحدى: باحدي Ox ٧ الحريف:
الحريف Ox || و⁵: + مثل Ox ٩ قباض بارد: قبا●●●رد Ox ١٥ الفنجنكشت: التـجكست Ox

---

101 *fāwaniyā* < *παιωνία* "peony", see LSLex 1289b. Ṭabarī's remark that the peony "has the size of a broad bean" must refer to the *seeds* of that plant; however, Galen, who is quoted here as an authority, talks of the *root* (*ῥίζα*) as a sympathetic remedy for (juvenile) epilepsy, see GalKü 11/859 (*Περὶ κράσεως καὶ δυνάμεως τῶν ἁπλῶν φαρμάκων*, cf. FiCG no. 78).

statement by Galen—cure epilepsy, to wit the drug called *fāwanīyā*[101] which has the size of a broad bean; and some are hung around the thigh to help against colic pain, like (dried) wolf's dung.

Aristotle already said: there are eight tastes—first, sweetness; then bitterness; then saltiness, sourness, acridness, sharpness, foulness, and greasiness.[102] These tastes clearly have astonishing effects on the body, which I will mention (next). §99

Most closely related to a balanced temperament is sweetness, because it comprises a mixture of one part of heat and one part of moisture; if either of these two powers increases or decreases, the taste (of the substance) is altered accordingly. As regards bitterness, it results from (a mixture of) heat and dryness. Anything salty is hot-dry (or) earthy, but lower in heat than bitter or acrid (stuff); saltiness is like sea (water), from which the sun has absorbed all thin (particles) such that only thick (matter) is left; a special property of saltiness is the effect that it dives deep into the body, dissolves moistures, and (thereby) protects an equilibrium. Anything astringent is cold-dry, witness fruits: they are hard and pungent in the beginning but, as they ripen under the heat of the sun and the rise of the moon, become (more) balanced and sweet, like grapes, apples, and pomegranates. Anything purely sweet is nourishing; anything in between sweet and bitter nourishes a little. Anything sour is mildly cold, cleans the channels of the body, and has a purifying (effect). Anything sharp is earthy, and has a drying (effect). Anything acrid is fiery and hot, and mollifies residues. Anything fatty is hot-moist, and softens and slackens the body without heating it up. Anything that increases (the production of) semen is hot-moist or hot and bloating; anything that interrupts (the production of) semen does so either through (a combination of) heat and dryness—like rue, mustard, hempseed, and monk's pepper—, or through (a combination of) coldness and dryness—like black poppy and lettuce seeds. §100

102 AriDA 96f. (*Περὶ ψυχῆς*): "As with the colours, so it is with the (different) kinds of taste (*χυμός*)—there are, firstly, simple tastes, which are opposites, (namely) the sweet and the bitter; next to these on one side the succulent, on the other the salty; and thirdly, intermediate between these, the pungent, the rough, the astringent, and the acid. These pretty much represent all the varieties of taste". As is to be expected with a lexical field such as this, the Greek and the Arabic terms, in themselves polysemous, do not always match up after transition; the Aristotelian terminology, for the record, is as follows: *γλυκύς, πικρός, λιπαρός, ἁλμυρός, δριμύς, αὐστηρός, στρυφνός, ὀξύς*.

§101 وأنا واصف قوى أشياء وما فيها من المنافع على انفرادها ومقتصِر عليها دون الأدوية المركبة، فإن فيها أشفية ومنافع كثيرة كافية على ما قالوا.

§102 فأولها الحنطة وهى حارة لينة جيدة الغذاء، والخشكار فيه حرارة يسيرة، والفطير غليظ ينفخ، وإن وضع الخمير على ورم مع دهن بنفسج أنضجه، وخاصية الحنطة أنها أغذى من سائر الحبوب، وأجود الخبز ما عجن وملك ملكا جيدا وطرح فيه من الملح والخمير قدر الحاجة وخبز في التنور ونضج ظاهره وباطنه وأحمده ما أكل بعد خَبْزه بيوم وما بعد ذلك إلى أن يصلب.

§103 الشعير بارد فيه يبس ونفْخ وماؤه غذاء ودواء يطفئ الحر وينقي الصدر.

§104 الأرز معتدل في الحر والبرد ييبّس البطن ويغذو غذاء محمودا ولا سيما إذا طبخ باللبن.

§105 الجاورس بارد يابس يغذو طبيخه ويحبس البطن.

§106 الباقلى اليابس وسط في برده ويبسه وإذا طبخ صار لدنا لينا وفي مائه تنقية وجلاء للصدر وفيه حرارة.

§107 الترمس حار يابس ينقي بمرارة فيه ويولد غلظا وإن شرب مع العسل والخل قتل الديدان وأيضا إذا ضمد به البطن.

§108 الحمص حار لين يزيد في المني وفي لبن النساء ويفتح سدد الكبد.

§109 اللوبياء دون الحمص في قوته وحرارته يولد غلظا.

---

٤ خاصية : خاصته Ox ١٢–١٣ وأيضا إذا ضمد به البطن ~ ṬabFir 375,13 f. : Ox non legitur ١٤ سدد الكبد ṬabFir 375,16 : Ox non legitur ١٥ اللوبياء دون ṬabFir 376,7 : Ox non legitur

I will now briefly describe the powers and benefits of certain simple substances, omitting compound preparations. These (simples) cover a sufficiently wide range of therapeutic applications, and (are given below) in line with what they[103] say. §101

First comes wheat, which is hot, emollient and very nourishing. (Its) bran contains little heat. (Its) unleavened dough is tough and bloats. (Its) leavened dough, when (mixed) with sweet violet oil and put on a tumour, ripens it. Wheat has the special property of being more nutritious than other (types of) grain. The best bread is obtained from a thoroughly kneaded (dough) into which salt and leaven have gone as required, and which has been baked in a pit-oven until its outside and its inside are well done; ideally, it is eaten within a day or so after baking, and before it gets stale. §102

Barley is cold, with some dryness and (a capacity to) bloat. Its water is (both) nutriment and remedy; it extinguishes heat and clears the chest. §103

Rice is balanced regarding heat and coldness. It dries the belly and provides laudable nourishment, especially when cooked with milk. §104

Millet is cold and dry. Its decoction nourishes, and astricts the belly. §105

Dried broad beans are moderately cold and dry. When cooked, they become tender and soft; their water, which contains (some) heat, has (a capacity to) clear and cleanse the chest. §106

Lupine (beans) are hot and dry. They purify (the body) through their bitterness, but (also) engender coarseness; when ingested with honey and vinegar, or when applied as a poultice to the belly, they kill (intestinal) worms. §107

Chickpeas are hot and emollient. They increase semen and breast milk, and open obstructions of the liver. §108

Cowpeas are less powerful and less hot than chickpeas. They engender coarseness. §109

---

103 That is the Greeks.

العدس بارد جاف يظلم البصر ويولد البواسير ويسكن الحر والدم وقشره حار حِرِّيف. §110

الحلبة حارة تهيج الباه وتلين الصدر وإذا وضعت على الظفر المتشنج أصلحته. §111

السمسم بطيء الانهضام حار لين يزيد في المني. §112

الخشخاش الأبيض بارد يابس يزيد في المني لدسومة فيه، فأما الأسود منه فإن أكثر منه مكثرٌ أضر به. §113

القرطم حار لين يسهل البطن. §114

الشهدانج حار يابس رديء للمعدة ويزيد في المني. §115

الماش بارد معتدل في اللين واليبس غير نفّاخ. §116

في البقول والكواميخ

أول البقول الخس وهو بارد رطب يولد دما جيدا وينوم وينفع من حرق النار إذا دق ووضع عليه وإن اكتحل من اللبن الذي يخرج منه نفع من العشى. §117

الهندباء دون الخس في البرودة والرطوبة يفتح سدد الكبد بمرارة فيه ويسكن اليرقان. §118

الجرجير حار لين يزيد في المني ويمرئ الطعام ويهيج الصداع. §119

الفجل حار وفيه يبس وتحليل الغلظ وإذا شرب ماء ورقه نقص اليرقان وأخرجه. §120

السلجم حار رطب ينفخ ويهيج الباه ويخرج البول ويزيد في المني. §121

٢ المتشنج : المتشيخ Ox   ١٢ اليرقان ṬabFir 377,18 : Ox non legitur   ١٤ الغلظ : + الذي فيه Ox   ١٥ يزيد في المني ṬabFir 378,3 : Ox non legitur

Lentils are cold and desiccative. They darken the sight and produce haemorrhoids, but they (also) appease heat and blood; their pods are hot and pungent. §110

Fenugreek is hot. It stimulates coitus and soothes the chest; when put on rugged fingernails, it mends them. §111

Sesame—slow to digest—is hot and emollient. It increases semen. §112

White poppy (seeds) are cold and dry; due to their fattiness, they increase semen. As for black poppy (seeds), they are harmful when taken in large quantities. §113

Safflower is hot and emollient. It purges the belly. §114

Hempseeds are hot and dry. They are bad for the stomach but increase semen. §115

Mung beans are cold, somewhat emollient and dry. They do not bloat. §116

On greenstuffs and vinegar-based pickles

First among greenstuffs comes lettuce, which is cold and moist. It generates good blood and promotes sleep; it is useful for burns when ground and put on (the area); and when the latex that exudes from its (stems) is applied as a collyrium, it helps against nightblindness. §117

Wild chicory is less cold and less moist than lettuce. It opens obstructions of the liver through its bitterness, and alleviates jaundice. §118

Rocket is hot and emollient. It increases semen and renders food (more) digestible, but it (also) excites headache. §119

Radishes are hot, with some dryness and (a capacity to) dissolve coarseness; the water of their leaves is drunk to reduce and (eventually) remove jaundice. §120

Turnips are hot and moist. They bloat, stimulate coitus, expel urine, and increase semen. §121

§122 الجزر حار رطب ينفخ ويهيج الباه ويخرج البول وينضج.

§123 الكرفس حار يابس يفتح سدد المعدة.

§124 النعنع حار لطيف يابس جيد للمعدة والكبد وإذا شرب من مائه مع ماء النمام مرارا كثيرة نفع من الفواق.

§125 الحبق النهري حار يابس ينفع من لسع الهوام.

§126 الكزبرة باردة يابسة إذا وضعت على البدن وإن أكثر منها أورثت خدرا.

§127 الهليون حار لين يزيد في المني.

§128 الباذرنجبويه حار لطيف ينفع شمه وأكله من خفقان القلب.

§129 الطرخون بارد يابس ثقيل.

§130 الفرفخ يبرد الحر الشديد إن شرب منه أو وضع على البدن ويقطع العطش.

§131 الكراث الشامي دون النبطي في الحر واليبس رديء للمعدة، والكراث النبطي حار يابس يصدع وينفع من البواسير وينزل البول، وفعل الشامي مثله.

§132 الثوم فائت في حره ويبسه والرطب منه أقل حرا ييبّس البدن وينفع من تغيير المياه في الأسفار ومن الرطوبة والبرودة ويوضع على لسع الأفاعي والعقارب وعضة الكلْب الكلِب فينفع ويسمى ترياق القرويين.

---

١ ينضج ṬabFir 380,21 : Ox non legitur ٣ جيد للمعدة والكبد وإذا شرب ṬabFir 378,5 : Ox non legitur ٣–٤ نفع من الفواق ṬabFir 378,6 : Ox non legitur ٥ الحبق النهري حار ~ṬabFir 378,6 f. : Ox non legitur ٦ باردة يابسة ṬabFir 378,8 : Ox non legitur ٨ الباذرنجبويه : البادرنجبوي Ox ١٠ الفرفخ : العرفج Ox ١٣ حرا ييبس : حر يبس Ox ١٥ القرويين : الفزوتين Ox

Carrots are hot and moist. They bloat, stimulate coitus, expel urine, and ripen (tumours). §122

Celery is hot and dry. It opens obstructions of the stomach. §123

Mint is hot, dry and lenient. It is good for stomach and liver; its water, when (mixed) with that of wild thyme and drunk in quick succession, is useful against hiccups. §124

Watermint is hot and dry. It is useful against vermin bites. §125

Coriander is cold and dry. If a lot of it is put on the body, it causes numbness. §126

Asparagus is hot and emollient. It increases semen. §127

Lemon balm is hot and lenient. Smelled or eaten, it is useful against palpitations of the heart. §128

Tarragon is cold, dry and heavy. §129

Purslane, when drunk or put on the body, cools down intense heat; it (also) quenches thirst. §130

Syrian leek is less hot and less dry than Nabatean (leek); it is bad for the stomach. Nabatean leek is hot and dry; it causes headache but is useful against haemorrhoids; it (also) unloads urine. The Syrian (variety) has the same effects. §131

Garlic is exceedingly hot and dry. Fresh garlic has less heat; it dries the body, serves to improve (the quality of) suspicious water when travelling, and (counters) dampness and coldness; it is useful (when) put on viper bites, on scorpion (stings) and on the morsus of a rabid dog. It is known by the name of peasants' theriac.[104] §132

---

104 *tiryāq al-qarawīyīn* "peasants' theriac" is a calque of Galen's made-up *ἀγροίκων θηριακή*, see GalKü 10/866 (*Θεραπευτικὴ μέθοδος*, cf. FiCG no. 69).

البصل فائت في حره رطب يزيد في المني ويطلى على الوجه ويذهب بالكلف وينفع من تغيير المياه. §133

الباذنجان حاريابس لمرارة فيه يولد إذا أكثر منه المرة السوداء. §134

الكرنب حاريابس رديء للبصر ومن أكل من لب قضبانه الرطب قوى على الشرب. §135

القنبيط حاريزيد في المني. §136

السلق حاريابس ينقي بملوحة فيه وينفع من سدد الكبد ويضر بالمعدة. §137

اللبلاب حاريابس يلين البطن ويخرج الصفراء. §138

الكبر حاريابس مفتح سدد الكبد والطحال بمرارته وحرافته وإن مضغ من قشره ووضع على الضرس الوَجِع نفعه ويقطر من مائه في الأذن فيقتل الدود. §139

الكشوث حاريابس لمرارة فيه وينفع لذلك الكبد والمعدة. §140

السرمق بارد رطب وبزره إذا شرب هيج القيء. §141

الخباز والبقلة اليمانية مثل السرمق يلين الصدر والبطن ويسكن الحر. §142

الإسفاناخ أحمد من البقلة اليمانية. §143

الباذروج حاريابس يجمد الدم ويظلم البصر وإذا استعط بمائه مع الكافور حبس الدم من الرعاف. §144

الصعتر حاريابس ينفع من برد المعدة ويلطف الرياح الغليظة. §145

٢ إذا أكثر منه المرة السوداء TabFir 379,7 : – Ox ٦ اللبلاب : اللباب Ox ٨ الضرس : الظرس Ox
١١ الخباز : الخناز Ox ١٣ استعط : استغط Ox

Onions are exceedingly hot and moist. They increase semen, make freckles disappear (when) spread over the face, and serve to improve (the quality of) suspicious water. § 133

Aubergines are hot, dry and somewhat bitter. They generate black bile when (consumed) excessively. § 134

Cabbage is hot and dry. It is bad for the sight; he who eats the pith of its fresh stalks, can drink (more wine). § 135

Cauliflower is hot. It increases semen. § 136

Beetroot is hot and dry. It purifies (the body) through its saltiness;[105] it is useful against obstructions of the liver but harmful to the stomach. § 137

Hyacinth beans are hot and dry. They soften the belly and expel yellow bile. § 138

Capers are hot and dry. They open, through their bitterness and acridness, obstructions of liver and spleen; their husks are useful when chewed or put on a painful molar; and their juice, (when) dripped into the ear, kills wigglers. § 139

Flax dodder is hot, dry and somewhat bitter. This is why it benefits liver and stomach. § 140

Orache is cold and moist. Its seeds, when ingested, provoke vomiting. § 141

Mallow and blite share (the properties of) orache. They (moreover) soothe the chest, (soften) the belly, and appease heat. § 142

Spinach is more laudable than blite. § 143

Basil is hot and dry. It clots the blood and darkens the sight; its water, when (mixed) with (that of) camphor and (then) snuffed, stops nosebleed. § 144

Savory is hot and dry. It is useful against coldness of the stomach and relieves tough winds. § 145

---

105 The sodium content of beetroot (*silq*) is in fact relatively low (c. 70 mg per 100 g).

الشبث حار يابس وإذا شرب مع ماء قد طبخ فيه يهيج القيء. §146

السذاب فائت في حره ويبسه ميبّس للرطوبات رديء للجماع وينفع ماؤه من خناق الرحم. §147

الخردل فائت في حره ويبسه نافع من كل داء من الرطوبة والريح الغليظة. §148

الحرف فائت في حره ويبسه رديء للمعدة وينفع من الرطوبات والرياح الغليظة وله خاصية في الذهاب إلى المواد الرديئة وإخراجها. §149

الخندقوقا حار يابس ينفع من برد المعدة والبلغم. §150

الراسن حار يابس نافع من البرد. §151

الحماض بارد يابس يدبغ المعدة وإن أكل مطبوخا أخرج الصفراء وينفع من سحج الأمعاء. §152

عنب الثعلب بارد يابس قبّاض نافع من ورم الكبد والمعدة وسددهما إن أكل نيئا أو مطبوخا. §153

الخيار بارد رطب رديء الانهضام يسكن الحرارة ويطفئ المرة الصفراء، وكذلك فعل القثاء إلا أنه أخف من الخيار، والإكثار منهما يولد رطوبة لزجة غليظة. §154

البطيخ حار رطب يجلو وينقي ويدر بزره البول والإكثار منه يحدث عفونات. §155

وكل كامخ يُعْمَل من شيء من هذه الأشياء فقوته مثل قوة ذلك الشيء. §156

في الفاكهة والثمار

عامة الفواكه تولد خلطا رديئا، والحلو منها حار، فأما النيء الحامض فبارد وقبّاض يحبس البطن. §157

---

١ يهيج : بهيج Ox ٢ ميبس : مييس Ox

Dill is hot and dry. Its decoction, when drunk, provokes vomiting. § 146

Rue is exceedingly hot and dry. It drains dampness but is bad for sexual intercourse; its water is useful against suffocation of the womb. § 147

Mustard is exceedingly hot and dry. It is useful against all damp-related diseases and against tough winds. § 148

Gardencress is exceedingly hot and dry. It is bad for the stomach but useful against dampness and tough winds; it has the special property of going to malevolent matters and evicting them. § 149

Sweet clover is hot and dry. It is useful against coldness of the stomach and against phlegm. § 150

Elecampane is hot and dry. It is useful against coldness. § 151

Sorrel is cold and dry. It fortifies the stomach; when eaten cooked, it expels yellow bile; it is (also) useful against bowel abrasion. § 152

Black nightshade is cold, dry and astringent. When eaten raw or cooked, it is useful against swellings and obstructions of liver and stomach. § 153

Cucumbers are cold and moist; they are badly digested but appease heat and stifle yellow bile. Serpent melons do the same, except that they are lighter (to digest) than cucumbers. (Consumed) in excess, they both generate thick, sticky moisture. § 154

Melons are hot and moist. They cleanse and clear (the body), and their seeds make the urine flow; (consumed) in excess, melons produce putrefying matters. § 155

The power of any (given) vinegar-based pickle is similar to the power of any of the (above) substances, insofar as they were used in its making. § 156

On fruits and fruity edibles

Fruit in general produces a bad (humoral) mixture. Sweet fruits are hot; as for unripe, sour (fruits), they are cold, astringent and constipate the belly. § 157

§158 ورأس الفواكه التين والعنب، فالتين حار يابس يلين الصدر والبطن وينفع من ورم الكبد والطحال ولا سيما إذا وضع على الطحال منقعا في الخل ويجلو الكلية والمثانة، والأبيض منه أخف من الأسود واليابس منه أحر من الرطب.

§159 والعنب أكثر غذاء من سائر الثمار وهو حار رطب، والأبيض أقل حرا ينفخ قليلا.

§160 الزبيب يلين الصدر إذا نزع عجمه وينشف البلة وعجمه بارد يابس يحبس البطن.

§161 الفرصاد وهو التوت بارد رطب يسهل البطن دون المشمش في الغاية وإذا شرب بعدهما شراب صلب أو سكنجبين لم يضر وما كان منه نيئا فهو بارد يابس.

§162 حب الآس بارد يابس فيه حر يسير يدبغ المعدة ويحبس البطن.

§163 التفاح كله بارد والحلو فيه شيء من حرارة والمر منه يولد خلطا معتدلا والحامض منه يضر بالعصب.

§164 الأترج قشره حار يابس وشحمه بارد رطب وحماضه بارد يابس يجلو وينقي الأوساخ الظاهرة والباطنة تنقية عجيبة وورقه حار هضوم وفي حبه دهن حار لين.

§165 الإجاص بارد يسكن الصفراء ويخرجها.

§166 التمر الهندي بارد ملين.

§167 النبق بارد يابس قبّاض.

§168 السفرجل بارد يابس قبّاض يقوي المعدة ويدبغها ويحبس البطن.

---

٧ يضر : يضره Ox ٨ الآس : الأس (sic) Ox || يابس + فيه بسر يديع المعدة ويحبس البطن Ox exenium || حر ṬabFir 382,4 : – Ox ١٣ الإجاص : الأجاص (sic) Ox

At the head of (all) fruits are figs and grapes. Figs are hot and dry: they soothe the chest and (soften) the belly; they are useful against swellings of liver and spleen, especially when soaked in vinegar and put on the latter organ; they (also) cleanse kidneys and bladder. White figs are lighter (to digest) than black ones; dried figs are hotter than fresh ones. §158

Grapes—more nourishing than other fruits—are hot and moist. White (grapes) have less heat and bloat a little. §159

Seedless raisins soothe the chest and dry up dampness; their seeds are cold and dry, and astrict the belly. §160

Firṣād, that is mulberries,[106] are cold and moist. They purge the belly to a lesser extent than apricots (do); there is no harm in following their (consumption) with a stiff wine or with oxymel. Raw mulberries are cold and dry. §161

Myrtle berries are cold and dry (but also) contain a little heat. They fortify the stomach and astrict the belly. §162

All apples are cold. Sweet (apples) contain some heat; bitter ones produce a balanced (humoral) mixture; and sour ones damage the nerves. §163

The peels of the citron are hot and dry; its pulp is cold and moist; its sap is cold and dry, and wonderfully clears and cleans (the body) from external as well as internal filth; its leaves are hot and (easily) digestible; and its seeds contain a mildly hot oil. §164

Plums are cold. They calm down yellow bile and (then) expel it. §165

Tamarinds are cold and emollient. §166

The fruits of the Christ-thorn are cold, dry and astringent. §167

Quinces are cold, dry and astringent. They strengthen and fortify the stomach, and constipate the belly. §168

---

106 *firṣād* and *tūt* "mulberries" are synonymous terms, see e.g. SchṬab 133 no. 177.

§169 الرمان الحلو منه حار لين جيد للصدر نافع من حرارة الكبد والحامض بارد لطيف يحبس البطن.

§170 ثمر العوسج بارد قبّاض.

§171 الموز معتدل في الحر رطب ملطخ للمعدة.

§172 الكمثرى معتدل في الحرارة والبرودة جيد للمعدة يحبس البطن.

§173 التمر في الجملة حار رطب ملين، البلح بارد قبّاض والحلو منه حار، الطلع يابس أبرد من البلح، الجمار بارد يابس.

§174 العناب حار رطب يسكن الدم نافع من السعال والربو.

§175 الغبيراء باردة يابسة قابضة تحبس البطن.

§176 الجوز حار رطب يستحيل إلى الصفراء سريعا.

§177 الجلوز أغذى من الجوز وأبطأ استمراء وإذا أكل مع التين أو علق على العضد نفع من لسع العقارب.

§178 اللوز الحلو منه معتدل في الحر ينقي الصدر والرئة إذا أكل بالعسل والنيء منه ألطف وينفع من سدد الكبد.

§179 حب الصنوبر لين معتدل في الحر نافع للباه ومن وجع الرئة والسعال.

§180 الحبة الخضراء حارة يابسة تنفع من البلغم واللقوة والفالج.

---

٣ الموز : التور Ox

---

107 *balaḥ* are green and small dates, or dates that never ripen, see LaLex 1/246b–c and DoSupp 1/108b; a parallel passage in the *Paradise of Wisdom* reads *busr* instead (ṬabFir

Sweet pomegranates are hot and emollient; they are good for the chest and useful against heat in the liver. Sour (pomegranates) are mildly cold and astrict the belly. §169

The fruits of the boxthorn are cold and astringent. §170

Bananas are somewhat hot and moist. They soil the stomach. §171

Pears are balanced regarding heat and coldness. They are good for the stomach and astrict the belly. §172

Dates, on the whole, are hot, moist and emollient. Unripe dates[107] are cold and astringent; sweet ones are hot; the spadix (of the date palm) is dry and colder than the unripe date; and palm pith is cold and dry. §173

Jujubes are hot and moist. They appease the blood and help against coughing and asthma. §174

The fruits of the service-tree are cold, dry and astringent. They constipate the belly. §175

Walnuts are hot and moist. They quickly transform into yellow bile. §176

Hazelnuts are more nourishing than walnuts and very slow to digest. When eaten with figs or hung on the upper arm, they are useful against scorpion stings. §177

Sweet almonds are somewhat hot. When eaten with honey, they clear chest and lungs. Unripe almonds are milder (in effect) and useful (moreover) against obstructions of the liver. §178

Pine nuts are emollient and somewhat hot. They help towards coitus, and against pain in the lungs and coughing. §179

Pistachios are hot and dry. They are useful against phlegm, facial paralysis, and hemiplegia. §180

383,6), which latter term denotes unripe dates, or dates that are turning from green to yellow and begin to ripen, see LaLex 1/202c and DoSupp 1/83b.

§181 جوز الهند حار رطب ثقيل يحبس البطن لقبض فيه ويزيد في المني ويخرج الديدان.

§182 الخوخ بارد رطب سريع العفونة، وكذلك المشمش رديء الخلط، ويتحرز من ضررهما إما بالقىء وإما بشرب شراب صلب صرف رقيق.

§183 الكمأة باردة رطبة ولها غذاء غليظ غير رديء.

§184 الفطر بارد رطب له غذاء رديء لزج وما ينبت منه تحت الزيتون فهو قاتل.

الرياحين

§185 الورد بارد يبرد حرارات الرأس والمعدة وكذلك ماؤه وييبس القروح.

§186 النيلوفر بارد رطب ينفع شمه من السهر الذي سببه الحرارة ومن حرارة الرأس.

§187 البنفسج بارد رطب ينفع المحرورين إن كان يابسا إذا شُمّ أو وضع على الرأس أو على أوجاع العين والمعدة الحارة ومن سكات الصبيان ويسهل البطن إذا شرب.

§188 اليبروح بارد يابس يورث السبات.

§189 المرزجوش حار يابس نافع من الصداع البارد يفتح السدد في الرأس ويسكن وجع الأذن وإذا وضع على لسع العقارب نفع من أوجاعها.

§190 النمام حار يابس يفتح سدد الرأس والأنف.

§191 السوسن حار يابس، لا سيما الأبيض منه، وهو نافع للعصب والمخ ويحلل الرطوبات والرياح الغليظة.

---

٢ بالقىء : بالنقي Ox ٥ لزج : لدح Ox ٧ حرارات : حرازات Ox || ييبس : يبس Ox ١٤ النمام : التمام Ox

Coconuts are hot, moist and heavy. They constipate the belly through their astringency; they increase semen and dislodge (intestinal) worms. § 181

Peaches are cold and moist; they quickly rot. The same goes for apricots, which have a noxious composition. One can take precautions against their (potential) damage either by vomiting (afterwards) or by drinking pure, thin, stiff wine. § 182

Truffles are cold and moist. They make for tough nourishment but are not malicious. § 183

Mushrooms are cold and moist. They make for bad, gluey nourishment, and those that grow under olive trees are lethal. § 184

Aromatic plants

Roses are cold. They cool down heat in the head and in the stomach, and dry out ulcers; their water (does) the same. § 185

Nenuphars are cold and moist. Smelling them helps against heat-induced insomnia and against heat in the head. § 186

Sweet violets are cold and moist. When smelled in dried form or put on the head, on the painful eye or on the hot stomach, they benefit fever sufferers; (they are useful moreover) against aphasia in children, and they purge the belly when ingested. § 187

The mandrake is cold and dry. It causes lethargy. § 188

Marjoram is hot and dry. It is useful against cold headache, opens obstructions in the head, relieves the ailing ear, and attends to painful scorpion stings when put on (the area). § 189

Wild thyme is hot and dry. It opens obstructions in the head and in the nose. § 190

Lilies are hot and dry, especially the white ones. They benefit nerves and brain, dissolve moistures, and (release) tough winds. § 191

§192 النرجس حار يابس ينفع من حرق النار إذا وضع عليه ويقلع البهق والبرش ويوضع أصله على الأورام الغليظة فينضجها وإن وضع على الجراحات أخرج النصل وغيره من البدن وسكن الصداع الذي سببه الحرارة.

§193 ورق الزعفران حار يابس يقبض قليلا وينوّم ويلبس القروح وإن وقع في الرحم نفع من تشنجها.

§194 الأقحوان حار يابس ينوّم وينفع من الناصور والأدرة التي تكون في الرطوبة إذا طلي عليها.

§195 الشاه إسفرم حار يابس إلا أنه إذا رش عليه الماء ثم وضع على البدن برده، والحماحم خير منه.

§196 الخيري حار يابس يفتح السدة التي في الرأس، ولا سيما الأصفر فإنه أفضل من الأحمر، والأبيض أقلهما نفعا.

§197 المرو حار يابس نافع لأصحاب الرطوبات وهو يصدع على السدة.

§198 المروماحوز حار يابس يفتح سدد الرأس وينفع من الخفقان البارد ومن أوجاع الأرحام للنساء الحبالى إذا شربْنه بالشراب.

§199 الآس بارد يبرد شمه حرارة الرأس وإذا صب من مائه على البدن حبس البطن وينفع من كانت طبيعته حارة رطبة.

§200 الشيح حار يابس.

§201 الغرب بارد يابس ينفع شربه من تقذف الدم.

§202 السرو حار يابس مقبض ويفني رطوبات البطن ويلبس القروح الرطبة والأدرة التي من الرطوبة.

---

٣ سببه supra lineam : اصله Ox in textu ٤ تشنجها : تشـخها Ox ٥ عليها : عليه Ox ٧ السدة : السدد Ox ١٦ الأدرة : الارزه Ox

Daffodils are hot and dry. They are useful against burns when put on (the area); they eradicate vitiligo and white leprosy; their roots, when put on thick tumours, ripen them, and when put on wounds, they extract arrowheads and other (objects) from the body; (daffodils also) soothe headache if its cause is heat. § 192

Saffron leaves are hot and dry. They astrict a little, induce sleep, and (are used to) cover ulcers; placed on the womb, they help against spasms. § 193

Daisies are hot and dry. They induce sleep and are useful against fistulas and hydrocele when smeared upon them. § 194

Sweet basil is hot and dry. However, if besprinkled with water and then put on the body, it has a cooling effect. Buglosses are better. § 195

Gillyflowers are hot and dry. They open an obstruction in the head—(this is true) especially for the yellow (kind), which is preferable to the red one; the white (kind) is less useful. § 196

Wild marjoram is hot and dry. It benefits those who are dominated by moisture, but it causes headache (when used) in a state of obstruction. § 197

Cat thyme is hot and dry. It opens obstructions in the head; it is (also) useful against cold palpitations; and when pregnant women drink it with wine, it eases pains in the womb. § 198

Myrtle is cold. (When) smelled, it cools down heat in the head; when some of its water is poured over the body, it astricts the belly; and it benefits those who have a hot-moist constitution. § 199

Wormwood is hot and dry. § 200

The weeping willow is cold and dry. (When) ingested, it is useful against spitting blood. § 201

The cypress is hot and dry. It astricts, annihilates dampness in the belly, and (is used to) cover moist ulcers and hydrocele. § 202

باذرنجبويه حار يابس نافع من برد المعدة وخفقان الفؤاد الذي من الصفراء إذا أكل على الريق وينفع من الفواق الذي سببه السوداء. §203

النسرين حار يابس ينفع أصحاب السوداء والبلغم ورياح الدماغ ووجع العصب وكذلك دهنه. §204

البابونج حار يابس ييبس الرطوبات التي في الدماغ إذا شُمّ. §205

الدهمشت وهو الغار حار يابس يفتت حصى المثانة نافع من الصداع البارد. §206

الأنبجات

الهليلج المربى يقوي المعدة وييبس الرطوبة ويلين البطن وينفع من الباسور وينفع أصحاب السوداء إذا أخذ على الريق، وكذلك الأملج المربى إلا أنه دونه. §207

الأترج المربى بطيء الهضم لا سيما إذا أكل بعد الطعام إلا أنه يستفيد من العسل قوة قاصلة ويربى بعسل وأفاويه. §208

الزنجبيل حار رطب يزيد في المني ويسخن البدن ويهضم وينفع أصحاب البرودات. §209

الشقاقل دون الزنجبيل في الحر وأقوى منه في الرطوبة يزيد في الماء. §210

القرع ينفع المحرورين ويستفيد من العسل حرا يسيرا. §211

الجزر حار رطب جيد للمعدة وينشف رطوبتها وينفع الكبد الباردة. §212

---

١ باذرنجبويه : بادرنجبوبي Ox

---

108 The concept of "winds in the brain" (*riyāḥ ad-dimāġ*) is clearly Indian rather than Greek—thus, Ayurvedic texts invariably feature separate sections on 'wind diseases' (*vātavyādhi*), and whilst the latter may be rooted in any bodily organ, they always produce pathological conditions associated with the *nervous system*; see JoIM 144 (and the literature quoted there).

Lemon balm is hot and dry. When eaten before food, it is useful against coldness of the stomach and against heart palpitations that are caused by yellow bile; it (also) helps against hiccups whose cause is black bile. § 203

Dog roses are hot and dry. They benefit those who are dominated by black bile and phlegm; (they are useful moreover against) winds in the brain and painful nerves;[108] their oil (does) the same. § 204

Chamomile is hot and dry. When smelled, it dries up dampness in the brain. § 205

Dahmašt, that is bay laurel,[109] is hot and dry. It crumbles bladder stones and is useful against cold headache. § 206

Preserves

Preserved myrobalans strengthen the stomach, dry up dampness, soften the belly, help against haemorrhoids and, when taken before food, benefit those who are dominated by black bile. Preserved emblics (do) the same, except that they are less (powerful). § 207

Preserved citrons are slow to digest, especially when they are eaten after a meal; they do, however, derive decisive power from honey, (which is why) they should be preserved in it, together with spices. § 208

Ginger is hot and moist. It increases semen, warms the body, promotes digestion, and benefits those who are overcome by coldness. § 209

Baby carrots are less hot than ginger but stronger in terms of moisture (content). They increase seminal fluid. § 210

Gourds are useful for fever sufferers; they derive a little heat from honey. § 211

Carrots are hot and moist. They are good for the stomach, dry up dampness, and (also) benefit the cold liver. § 212

---

109 *dahmašt* (< *dah*[*a*]*mast*) is the Persian equivalent of Arabic *ġār* "bay laurel", see SchṬab 199 no. 305.

§213 السفرجل نافع للكبد يحبس البطن.

§214 الجزيْر المربى أمرأ من النيء جيد للجماع يستفيد بالعسل حرارة وقلة نفخه.

المياه

§215 الماء فيه حياة كل حيوان ونبت، وأفضل المياه على ما قال بقراط الحكيم ما كان منها أبيض صافيا خفيفا طيب الريح مما يسخن سريعا ويبرد سريعا فإن سرعة استحالته تدل على لطافته فأما البطيء الاستحالة فإنه يدل على غلظه، والمياه المالحة الثقيلة تيبس البطن، ومياه الثلوج والجليد رديئة ثقيلة، ومياه البطائح والسباخ حارة غليظة في الصيف لركودها ودوام طلوع الشمس عليها فهى تولد المرة الصفراء وتعظم الطحال والكبد، ومياه العيون التي تنبع من أرضين حارة رديئة لأن فيها أجزاء من تلك الأرضين، ومَن أدمن شرب الماء الحار ليّن عصبه وهيّج الرعاف وأرخى البدن وفرّق الحرارة الغريزية، ومَن أدمن الاغتسال بالماء البارد اسودّت بشرته وهيّج الكزاز والنافض لأنه يحبس الرطوبات على البدن ويعفنها وهو رديء للعصب والعظم، والاغتسال بالماء الحار نافع لكل عضو بارد إذا اقتصد فيه ويحلل الرطوبات ويلين البدن ويفتح منافذه وإن أكثر منه أضر بالعصب والعظم والمخاخ لأنه يحدث فيها التليين والرطوبات ما يغير خلقتها الطبيعية والإفراط منه يحدث الغشى، فأما الماء البارد فإن الإفراط فيه يحدث من هذه الأجزاء يبسا مفرطا وصلابة مخرجة لها من خلقتها الطبيعية.

٦ المياه : المياة Ox ٧ البطائح : البطـايح Ox ٨ تنبع : تقع Ox || لأن : الا ان Ox ١٤ الغشى : العشا Ox

110 The form *ǧuzair* (dim. of *ǧazar* "carrot"), though strongly suggested by the manuscript reading (حزير), is hard to back up; my translation "baby carrots" therefore remains somewhat tentative (the German translator of the text also decided on "Möhren", see ṬabHT 106).

Quinces are useful for the liver and astrict the belly. § 213

Preserved baby carrots[110] are more digestible than raw ones. They are good for sexual intercourse; they derive (additional) heat from honey, and (through it become) less bloating. § 214

Waters

The life of all animals and plants (depends) on water. The best water, according to what Hippocrates the sage said, is transparent, pure, light, sweet-smelling, quick to warm up and quick to cool down[111]—for rapid changeability indicates fineness, whereas slow changeability indicates crudeness. Saline, heavy water desiccates the belly. Snow water and ice water are heavy and noxious. Water from valley bottoms and marshlands is hot and crude in summer, because it stagnates and because the sun shines on it relentlessly; this (water) generates yellow bile and enlarges spleen and liver. Spring water that emerges from warm soils is noxious, because it contains earthy particles. Habitual drinking of warm water softens the nerves, excites nosebleed, slackens the body, and scatters the innate heat. Habitual washing with cold water darkens the skin and provokes trembling and shivering fits, because it locks away and (then) putrefies bodily fluids, which is bad (also) for the nerves and the bones. Washing with warm water, done in moderation, benefits all cold organs, dissolves moistures, softens the body, and opens its pores; too much of it (however) damages the nerves, bones and (spinal) marrows by making them lax and humid, and thereby altering their natural condition; and (when) done excessively, it may (even) cause fainting. Back to cold water: excessive (washing) with it makes these organs[112] very dry and stiff, well beyond their natural condition. § 215

111 HippLi 2/32 f. = HippHe² 62 (*Περὶ ἀέρων ὑδάτων τόπων*, cf. FiCH no. 2): “Les eaux de pluie sont les plus légères, les plus douces, les plus ténues, les plus limpides” and HippLi 4/542 f. (*Ἀφορισμοί*, cf. FiCH no. 13): “L’eau qui s’échauffe promptement et se refroidit promptement est la plus légère”. Much of what follows in this paragraph are abridged and paraphrased statements made also by Hippocrates on the subject of ‘waters’, see HippLi 2/26–37 = HippHe² 60–63 (*Περὶ ἀέρων* …); in § 94, Ṭabarī already reported some *Indian* teachings about water.

112 Lit. “parts” (*ağzā’*).

في منافع أعضاء الحيوان

§216 قال بعض الفلاسفة إن شعر الإنسان إذا بل بالخل ووضع على عضة الكلْب الكلِب نفع من ساعته، وبزاق الإنسان ينفع من لدغ الهوام، ولبن النساء إذا شرب مع الشراب أو العسل فتت حصى المثانة، وبول الإنسان ينفع من جميع الهوام القاتلة ومن عضة الكلْب الكلِب إذا صب عليها، وإذا علق عظم الإنسان على من به حمى الربع نفعه نفعا بينا، وقد قال جالينوس إن زبل الإنسان إذا يبس وعجن بالعسل وطلى على الحلق نفع من الخناق وكذلك إن شرب منه.

§217 وإن أخذ حافر برذون وأحرق وأخذ رماده ووضع على الخنازير معجونا بزيت عفنها، وإن دخنت المرأة الحبلى بروث البراذين أخرج الجنين حيا كان أو ميتا وإذا يبس الروث وذر منه على الجراحات حبس الدم والرعاف، وإذا أخذ وسخ من أذن بغلة وعلق على المرأة في بندقة منع من الولادة.

§218 وإن صب دم البقر على الجراحة حبس الدم، وإن أخذت مرارة البقر وخلطت مع نطرون وشحم الحنظل وعسل وطليت به المقعدة أسهل البطن وقال دياسقوريدوس إنْ قطر من مرارة البقر في

---

٣ فتت : قتت Ox ٤ عليها : عليه Ox ١١ الجراحة : الخراجة Ox ١٢ دياسقوريدوس : دياسقوريديوس Ox ١٢–١٤٤.١ في الأذن ṬabFir 422,7 : – Ox

---

113 Whilst the vast bulk of this chapter is dedicated to 'animal parts' (§§ 217–243), it begins with organs and products obtained from humans (§ 216) and ends with ink, lime and certain clays (§§ 244–246). Ṭabarī rarely cites or even implies his sources for the chapter in hand, but it is clear that a great deal of the material traces back to either Dioscorides' *Περὶ ὕλης ἰατρικῆς* (books 2 and 5 [DioWe 1/121–165 and 3/103–108]) or Galen's *Περὶ κράσεως καὶ δυνάμεως τῶν ἁπλῶν φαρμάκων* (books 9–11 [GalKü 12/159–377, cf. FiCG no. 78]), among others (cf. UllMed 108). The largely sympathetic prescriptions promoted here by Ṭabarī were soon after taken up to form the basis of a new genre of Arabic medico-pharmaceutical literature, dedicated exclusively to occult properties (*ḫawāṣṣ* or simply *manāfiʿ*) of human, animal and mineral substances and products; it is therefore no surprise that almost all of Ṭabarī's material on that subject, both in the present text and in the *Paradise of Wisdom*, is echoed and amplified by his younger contemporary ʿĪsā ibn ʿAlī, an early pioneer of the genre in its Arabic garb, see RaggAP xi–xxii. Due to the ubiquity and opacity of these ideas, I have occupied myself with source verifications only in those rare cases where Ṭabarī himself provides a more or

On the useful properties of animal parts[113]

One of the philosophers said: human hair, when wetted with vinegar and put on the bite of a rabid dog, is useful instantly.[114] Human saliva is useful against vermin bites.[115] Breast milk, when drunk together with wine or honey, crumbles bladder stones. Human urine is useful against all deadly creepers and against the bite of a rabid dog when poured over (the area). If a human bone is hung on someone who suffers from quartan fever, it clearly benefits him. Galen already said: human feces, when dried, kneaded with honey and painted on the throat, are useful against quinsy; the same (is true) when some of it is ingested.[116] §216

If one takes the hoof of a workhorse, burns it, kneads the ashes with olive oil and puts that on scrofula, it breaks them down. If a pregnant woman fumigates herself with horse dung, it evicts the foetus, dead or alive; if the dried dung is sprinkled on wounds, it stanches the blood and (also) stops nosebleed. If one takes the earwax of a female mule and hangs it, in a hazelnut (shell), upon a woman, it prevents (early) childbirth. §217

If cattle blood is poured over a wound, it stanches the blood. If one takes cattle gall, mixes it with natron, colocynth pulp and honey, and smears that on the anus, it purges the belly. Dioscorides said: cattle gall, when dripped §218

---

less explicit hint; this procedure is also justified by the fact that, as far as Arabic literature is concerned, there simply are no extant textual prototypes.

114 Instead of "one of the philosophers" (*baʿḍ al-falāsifa*), the parallel passage in the *Paradise of Wisdom* has "the philosopher Anaximenes" (اطروميس lege انكزيمنس < Ἀναξιμένης), see ṬabFir 420,4; Siddiqi's emendation (loc.cit. note ٢) ايكزوميس scil. Axominos (?) is wrong. It is tempting though hardly justifiable to read the Greek name 'Xenocrates' (of Aphrodisias, cf. note 116 below) into the Arabic graphograms. The statement itself is also found RaggAP 8 f. no. 1.8 (Arabic tradition).

115 This statement is explicitly (and wrongly) attributed, in the *Paradise of Wisdom*, to Dioscorides, see ṬabFir 420,7; it is found, for example, RaggAP 16 f. no. 1.27 (Arabic tradition).

116 Galen does, in fact, strongly disapprove of the therapeutic use of human feces, even though, as he says, 'Xenocrates' recommended them to be smeared into the mouth and onto the throat, and even to be ingested (GalKü 12/249; the reference here is to Xenocrates of Aphrodisias' infamous, organo-therapeutic work *Περὶ τῆς ἀπὸ τῶν ζώων ὠφελείας*). Ṭabarī's 'quotation' can, however, be traced to an interpolated passage in Dioscorides' chef-d'œuvre (DioWe 1/164,5 [app.] with BeDio 193), where human feces, mixed with honey and painted on the throat, are registered as a 'secret' remedy against tonsilitis; as is to be expected, the very same prescription resurfaces RaggAP 8 f. no. 1.5.

الأذن نفعت من الدوي والطنين وإن وضعت المرارة عليها مع دهن ورد وقطران سكن وجعها وإن غسلت بها الرأس نفعت من الحزازة، وإن أخذ من خصى العجاجيل ويبست وشرب من نحاتتها أنعظت وقويت على الباه، وأخثاء البقر توضع على الأورام الغليظة فتحللها وإن وضعت الأخثاء على رجل من به النقرس مع شيء من رماد ودهن زيت نفعه وإن أحرقت الأخثاء ووضع منها في المنخرين مع الخل حبس الرعاف وهو نافع من جميع السمام.

§219 وإن أخذت شعرة من ذنب الحمار إذا هو نزا على الأتان وعلقت على البدن أنعظت، ولبن الأتن ينفع من الأدوية القاتلة ومن قروح الأمعاء والزحير.

§220 ولحوم الكباش تنفع من شرب الذراريح، ويؤكل كبدها مشويا فيحبس البطن، ويقطر من مرارتها في الأذن مع شيء من ماء وعسل فيسكن الوجع.

§221 وإن شوى كبد المعز وقطر من مائه في عين من به عشى نفعه، وإن أخذ من جلده ساعة يسلخ ويوضع على لسع الحيات نفع، وينفع لبنها من قروح الأمعاء، ويوضع من بعرها على لسع الهوام وعض السباع فينفع وإن سحق بعرها وعجن بالعسل وطلى به البدن نفع من وجع المفاصل.

§222 بول الكلاب يطلى على الثآليل فيقلعها، قال دياسقوريدس إنْ أكل من قد عضه الكَلْب الكَلِب كبد الكلب نفعه، ومن علق نابا من أنياب الكَلْب الكَلِب على ساعده لم يضره الكَلْب الكَلِب.

§223 وإن غسل الرأس ببول البقر منع الحزاز والقروح وإن قطر منه في الأنف نفع من قروحه.

§224 ولبن اللقاح ينفع من فساد الكبد وينفع الجماع، وإن أحرق شعره وذر في الأنف حبس الرعاف.

---

١ نفعت : نفع Ox ٢ يبست : يبس Ox ٣ نحاتتها أنعظت وقويت : ىحاىىه اىعض وقوى Ox || توضع : يوضع Ox ٦ ولبن : أنعظت ولين انعضت Ox ٨ تنفع : ينفع Ox

---

117 DioWe 1/160 no. 78 (*Περὶ ὕλης ἰατρικῆς*): "Ox gall heals [...] purulent and cracked ears when dripped into them together with goat milk or breast milk, and it is also useful against tinnitus (when mixed) with leek juice; [...] smeared on dandruff together with saltpeter (*νίτρον*) or Cimolian earth (*γῆ Κιμωλία*, cf. note 124 below), it is excellent".

into the ear, is useful against tinnituses; when put on the ear together with rose oil and tar, it alleviates pain; and when used to wash the head, it helps against scurf.[117] If one takes calf testicles, dries them and ingests a few slivers, they cause an erection and enable coitus. Cattle dung, put on thick tumours, dissolves them; if the dung is put, together with a little ash and olive oil, on a man who suffers from gout, it benefits him; if the dung is burned and some of it is put, together with vinegar, into the nostrils, it stops nosebleed; and it is useful against all poisons.

If one takes a hair from the tail of a donkey, just as it mounts the female, and hangs that on the body, it causes an erection. Donkey milk is useful against (potentially) lethal drugs, against bowel ulcers, and (against) dysentery. §219

The meat of rams is useful against ingested cantharides. (Sheep) liver, eaten roasted, astricts the belly. Sheep gall, dripped into the ear together with some water and honey, alleviates pain. §220

If one roasts the liver of a goat and drips some of its juice into the eye of somebody who suffers from nightblindness, it benefits him. If one takes a goat hide, straight after flaying, and puts it on a snake bite, it helps. Goat milk is useful against bowel ulcers. Goat dung, put on the stings of vermin or on the bites of predators, is useful; and when the dung is pounded, (then) kneaded with honey and smeared on the body, it is useful against pain in the joints. §221

The urine of dogs, smeared on warts, extirpates them. Dioscorides said: if he who has been bitten by a rabid dog eats the liver of that dog, it will benefit him; and he who hangs one of the fangs of a rabid dog on his forearm, will suffer no harm from such an animal.[118] §222

If one washes the head with cattle urine, it prevents scurf and ulcers; and if some of it is dripped into the nose, it is useful against sores. §223

The milk of she-camels is useful against liver corruption and benefits sexual intercourse. Camel hair, when burned and sprinkled into the nose, stops bleeding. §224

---

118 DioWe 1/135 no. 47: "It is said that those who eat the roasted liver of the mad dog that has bitten them, will be protected from hydrophobia; as a precaution, they also make use of that dog's tooth by putting it in a bag (κύστις) which they fix to the arm".

وإن أخذت عين الذئب اليمنى وجففت وعلقت على الطفل لم يفزع، وكذلك أسنانه أو جلده. §225

وإن أخذ من أسنان الثعالب اليمنى وعلقت على الأذن اليمنى سكن وجعها وإن علقت سنه اليسرى §226 نفعت الأذن اليسرى.

وقال دياسقوريدس إنْ وضعت الفأرة مشقوقة على لسع العقرب نفعت نفعا بينا وإن وضعتَها §227 مشقوقة على الجراحة أخرجت النصول والشوك.

وإن دخن أطراف البدن بشعر الأرانب لم يخف البرد، وإن شربت المرأة خصية الأرنب أو §228 إنفحته رزقت ولدا ذكرا وإن شرب منها قدر باقلاة مع شراب صلب نفع من الربع، وإن وضعت إنفحة أرنب مع الخطمي والزيت على البدن أخرج النصول والقصب وإن سقى الصبيان من إنفحة الأرنب نفع من الفزع.

وإن أخذ جلد القنفذ وسحق وخلط بعسل وطلى على داء الثعلب أنبت الشعر، ويكتحل من مرارته §229 فينفع من بياض العين وإن شرب من مرارته نفع من الجذام والزحير والسل.

وإن وضعتَ الديك والدجاج مشقوقا على لسع الحيات وعض السباع نفع، وكذلك إن شرب من §230 دماغه وإن عجن منه بغبار الرحى وشرب منه قدر باقلاة نفع من نفث الدم، وإن دخن بزبله نفع من وجع الأسنان وإن أخذ زبل الدجاج وشرب منه مع العسل نفع من القولنج.

وإن خلط شحم البط بالشمع وطلى على الوجه نقاه وجلاه. §231

---

٢ علقت[1] : علق Ox ٣ نفعت : نفع Ox ٤ نفعت : نفع Ox ٨ النصول : الفضول Ox ١٣ نفث : سعث Ox

If one takes the right eye of a wolf, dries it and hangs it upon an infant, the latter shall have no fear. The same (goes for) the fangs and the hide of a wolf. § 225

If one takes the right fang of a fox and hangs it on the right ear, it alleviates pain; the left fang is useful when hung on the left ear. § 226

Dioscorides said: if one puts a split mouse on the sting of a scorpion, it is clearly useful;[119] and if you put a split mouse on a wound, it extracts arrow-heads and thorns. § 227

If one fumigates the limbs of the body with hare's hair, he need not fear the cold. If a woman ingests the testicles or the rennet of a hare, she will be blessed with a male child; and if a quantity of one broad bean of that is drunk with stiff wine, it is useful against quartan (fever). If hare's rennet is put, together with marshmallow and olive oil, on the body, it extracts arrow-heads and splinters; and if some of it is given to children in a drink, it helps against fearfulness. § 228

If one takes the skin of a hedgehog, pounds it, (then) mixes it with honey and smears it on (an area of) alopecia, it regrows hair. Hedgehog gall, applied as a collyrium, is useful against leucoma; when ingested, it helps against leprosy, dysentery, and consumption. § 229

If you put a split cock or hen on the bite of a snake or on that of (another) predator, it is useful. The same (is true) if one ingests some (chicken) brain; and if the latter is kneaded with the dust of millstones and ingested in a quantity of one broad bean, it helps against spitting of blood. Fumigation with chicken droppings is useful against toothache; and when ingested together with honey, these droppings help against colic.[120] § 230

If one mixes duck fat with beeswax and paints that on the face, it clears and cleanses. § 231

119 DioWe 1/143 no. 69: "It is generally said that split mice (*μύες*) are usefully applied to scorpion stings".

120 A similar statement is explicitly (and wrongly) attributed, in the *Paradise of Wisdom*, to Dioscorides, see ṬabFir 433,2f.

§232 فأما الحمام فإن دمه ينفع من الرعاف إذا قطر منه في الأنف وإذا طلى على العين نفع من العشى، وإن أخذ قدر ملعقة أو ملعقتين من زبل حمامة قد ألقمت أياما من دقيق الباقلى وحده نفع من أسر البول وحصى المثانة.

§233 وإن أخذ من كبد الحجل قدر مثقال نفع من الصرع.

§234 وقد تنفع مرارة الحجل ومرارة الكركي ومرارة السمك وعامة المرارات من ابتداء الماء في العين وفيها كلها جلاء للعين إذا اكتحل منها.

§235 ولحوم العصافير وبيضها تزيد في الباه، وإن أخذ زبلها وذيف بلعاب الناس وطلى على الثآليل نفعها.

§236 والذباب ينفع من أوجاع العين وإن أحرق الذباب وطلى بالعسل على داء الثعلب أنبت الشعر.

§237 وإن أخذ من الجراد الطوال الأرجل وعلق على من به حمى الغب نفع.

§238 والجندبادستر يسخن الأعضاء الباردة إذا شرب منه وينفع من اللقوة وبرودة الدماغ وينفع من النسيان والخفقان.

§239 وإن أخذ من السرطان وسحق ووضع على البدن أخرج النصول والشوك وإن شدخ السرطان شدخا ووضع على لسع العقرب نفع من ذلك ومن لسع الأفاعي والحيات.

§240 وإن أحرق الضفدع وسحق ونفخ في الأنف حبس الرعاف وإن خلط رماده بالزيت وطلى على داء الثعلب نفع نفعا بينا، وقال جالينوس إنْ أخذتَ الضفدع ووضعتَه على لسع الحية والعقرب نفع من ذلك.

---

١ في الأنف : − Ox   ٧ تزيد : يزيد Ox   ٨ داء : ذا Ox   ١٥ وضعته : وضعتها Ox   ١٦ نفع : نفعت Ox

---

121 Galen has, in fact, only the following to say about frogs (*βατράχοι*): "At ranarum ustarum cinerem sanguinis eruptioni mederi referunt inspersum: ceterum cum liquida pice sanare alopecias", see GalKü 12/362 (*Περὶ κράσεως καὶ δυνάμεως τῶν ἁπλῶν φαρμάκων*, cf.

As regards pigeons, their blood is useful against bleeding when dripped into the nose; and when painted on the eye, it helps against nightblindness. If one swallows a quantity of one or two spoonfuls from the droppings of a pigeon that has been fed nothing but broad bean meal for a few days, (these droppings) are useful against ischuria and bladder stones. § 232

If one swallows a quantity of one *miṯqāl* of partridge liver, it is useful against epilepsy. § 233

The gall of the partridge, the crane, the fish—(in fact) galls generally—may be useful in the early stages of cataract, (just as) they all have (a capacity to) clear the eye when applied as a collyrium. § 234

The meat of sparrows as well as their eggs increase (the desire for) coitus. If one takes their droppings, mixes them with human saliva and smears that on warts, it is useful. § 235

Flies are useful against eye pain; if one burns some flies and (then) smears them with honey on (an area of) alopecia, they make hair grow. § 236

If one takes long-legged locusts and hangs them on someone who suffers from tertian fever, they are useful. § 237

Castoreum warms the cold organs when some of it is ingested; it is useful against facial paralysis and coldness of the brain; it (also) helps against forgetfulness and palpitations. § 238

If one takes a crab, pounds it and puts (the pulp) on the body, it extracts arrowheads and thorns; if the crab is broken apart and put on the sting of a scorpion, it is useful against that, (just as it helps) against the bites of vipers and (other) snakes. § 239

If one burns a frog, pounds it and blows (the ashes) into the nose, they stop bleeding; if the ashes are mixed with olive oil and smeared on (an area of) alopecia, they are clearly useful. Galen said: if you take a frog and put it on the bite of a snake or (the sting of) a scorpion, it is useful against that.[121] § 240

FiCG no. 78); the 'quotation' therefore seems to relate rather to the first part of this paragraph.

§241 وإن أخذ من الحية نابها الأيسر وعلق على الأذن سكن وجع الأسنان، وإن علق قلبها على من به حمى الربع نفعه، وإن علق قرنها على من به حمى الغب نفعه، وسلخ الحية إذا جفف وسحق بشراب واكتحل به نفع البصر.

§242 العقرب إن أكل منها مشوية أو مسحوقة نفعت من لسعها ويعمل منها دواء يفتت الحصى من المثانة.

§243 وسام أبرص الذي يكون في البساتين إن شق ووضع على البدن أخرج النصول والقصب، وإن علق قلبه على النساء منع الإسقاط ونفع منه.

§244 المداد إذا طلي به موضع الحرق نفع نفعا بينا.

§245 وكذلك النورة إذا غسلت مرارا وصيرت مع دهن ورد وورق السلق وبياض البيض نفعت من حرق النار ومن الحمرة نفعا بينا.

§246 الطين الأحمر المختوم يحبس الدم وينفع من الخلفة، وكذلك الأرمني، والطين الحريطلى على لسع الزنابير بالخل فيسكن الوجع.

§247 وفي هذه الأدوية المفردة جُلُّ ما يوصف لمنافع الأبدان وأنواع العلل، وأنا ذاكر أدوية خفيفة إن شاء الله تعالى.

§248 فأولها دواء يقوي المعدة ويحرك الشهوة وينشف البلة من المعدة، يؤخذ من الإهليلج الكابلي الجيد جزء ومن الزنجبيل الصيني جزء ويدق وينخل ويعجن بعسل منزوع الرغوة، الشربة منه درهمين إلى ثلاثة على الشبع والريق.

---

٤ نفعت : نفع Ox || منها[2] : منه Ox ٩ نفعت : نفع Ox

If one takes the left fang of a snake and hangs it on the ear, it alleviates toothache. If a snake's heart is hung on someone who suffers from quartan fever, it benefits him. If a snake's horn is hung on someone who suffers from tertian fever, it benefits him. Snakeskin, when dried, pounded with wine and applied as a collyrium, is useful for the sight. §241

The scorpion, eaten roasted or pounded, is useful against its (own) sting; one (also) prepares from it a drug that crumbles bladder stones.[122] §242

The wall gecko, which lives in gardens, extracts arrowheads and splinters when split and put on the body. If one hangs its heart upon a woman, it serves to prevent miscarriage. §243

Ink, when smeared on the area of a burn, is clearly useful. §244

The same (goes for) lime: when washed several times and brought together with rose oil, beet leaves and egg white, it is clearly useful against burns and against erysipelas. §245

Red sealing bole stanches the blood and helps against diarrhoea. Armenian (bole does) the same. Pure clay, (when mixed) with vinegar and smeared on hornet stings, soothes the pain. §246

These simple drugs cover most of what is (generally) prescribed to advance physical well-being and (to counter) pathological conditions. And now I shall mention a few convenient (compound) drugs—God the Sublime willing. §247

First, a remedy which strengthens the stomach, stirs the appetite, and dries up gastric fluid: take one part of good chebulic myrobalans and one part of Chinese ginger, grind that, strain it, and knead it with combless honey; the dose is two to three *dirham* before or after food. §248

---

122 Galen has this 'drug': "The scorpion, dried and eaten with bread, crumbles bladder stones", see GalKü 14/242 = BouTP 45 (*Περὶ θηριακῆς πρὸς Πίσωνα*).

§249 دواء آخر يسهل البطن ويقوي المعدة ويشرب على الشبع والريق، يؤخذ من المصطكى زنة خمسة دراهم ويسحق بالماء حتى يصير مثل الكحل ويخلط بزنة عشرة دراهم سليماني، ويشرب منه كل يوم وزن درهم إلى أربعة دراهم فإنه يخرج كل فضل في البدن.

§250 دواء مسهل يخرج الفضول ويشرب على الشبع والريق، يؤخذ هليلج أصفر وهليلج أسود من كل واحد زنة خمسة دراهم ومن الكثيراء والمصطكى والورد الأحمر والزعفران من كل واحد زنة درهمين يدق وينخل ويعجن بماء الهندباء، الشربة منه زنة درهمين إلى ثلاثة دراهم.

§251 دواء ينفع من الحزاز في الرأس والإبرية، يؤخذ من مرارة الثور ومن قيموليا ويغسل به الرأس أو بحمص مدقوق معجون بخل خمر مع الخطمي.

§252 صفة الترياق الذي يؤخذ من أربعة أخلاط نافع من لسع الحيات وسائر الهوام والأخلاط الباردة الرديئة ويعمل على الترياق الكبير، يؤخذ من الجعدة والجنطيانا الرومي والبارزد وأصل الكبر أجزاء سواء تدق وتنخل بحريرة وتعجن بدهن اللوز أو الحل، الشربة منه حبة إلى دانقين بسمن بقر أو بدهن الحل.

§253 خضاب مجرب فائق قد جربتُه أحيانا، يؤخذ من الخطر الجيد زنة أربعين درهما ومن الحناء الزبداني الجيد زنة خمسة يخلطان معا ويسحقان في الهاون سحقا شديدا حتى يحمى الهاون ثم يلت ذلك

---

٤ هليلج[1] : هليليج Ox ٧ أو : و Ox ١٠ البارزد : الفازرد Ox || الكبر : الكبير Ox ١٢ الحل : بـالحل Ox ١٣ الزبداني : الربدانى Ox

---

123 This is a hard, crystallized candy made from white cane sugar, see for more details NaACK 601f. (*Sulaimānī*); the German translator of the text gives "Ätzsublimat" (ṬabHT 117), which is nonsense.

124 (*ṭīn*) *Qīmūliyā* < (*γῆ*) *Κιμωλία* "Cimolian (earth)" is a white clay from Cimolus, an island in the Cyclades; for this and other kinds of medicinal earths see the discussion SchṬab 293f. no. 476 and 417ff. no. 661.

125 Cf. ṬabFir 451,19–23 for a different version of this drug; another early example is found in Sābūr's dispensatory under the generic name "the *four ingredients* theriac" (*tiryāq al-arbaʿa al-adwiya*), see SābAq = SābDis no. 2.

126 *tiryāq kabīr* "great theriac", later also known as *al-fārūq* "the one that makes a difference (between life and death)", is the name of an ancient 'antidote' made famous

Another remedy which purges the belly, strengthens the stomach, and which is ingested before or after food: take an amount of five *dirham* of mastic, pound it with water until it becomes like a collyrium, (then) mix it with an amount of ten *dirham* of (ground) Solomon sugar,[123] and swallow that in a dose of one to four *dirham* daily—it does away with any residual matter in the body. §249

A purgative remedy which expels residues, and which is ingested before or after food: take an amount of five *dirham* each of yellow myrobalans and black myrobalans, and an amount of two *dirham* each of tragacanth, mastic, red roses and saffron, grind that, strain it, and knead it with wild chicory water; the (right) dose has a weight of two to three *dirham*. §250

A remedy which is useful against scurf on the head and eczema: take ox-gall and Cimolian (earth),[124] and wash the head with it; or (take) ground chickpeas kneaded together with wine vinegar and marshmallow. §251

Instruction (on how to make) the theriac that is composed of four ingredients[125]—it is useful against the bites of snakes and other vermin, (it counters) cold, noxious (humoral) mixtures, and it works in analogy to the *great theriac*:[126] take equal parts of germander, Greek gentian, galbanum and caper roots, grind that, strain it through a cloth of silk, and knead it with almond oil or sesame oil; the dose is one *ḥabba* up to two *dāniq*, with ghee or with sesame oil. §252

An excellent, well-tried dye which I myself have occasionally used: take an amount of forty *dirham* of good indigo (leaves) and an amount of five (*dirham*) of good henna (leaves) from Zabadān,[127] mix that together, and pound it so vigorously that the mortar becomes hot; then moisten it with §253

---

originally by Galen, see GalKü 14/82–89 (*Περὶ ἀντιδότων*) (followed by two variant transmissions); cf. ṬabFir 449,9–450,20 for an early Arabic version. For the preparation of this complicated drug in 'canonical' Arabic pharmacy see ISQā 3/312,3–20 (preceded by its therapeutic properties and followed by two variant transmissions [Arabic only]); cf. also KaAnt 479 ff. and 482–490 (with the literature quoted there, esp. WatStu *passim*).

127 Zabadān(ī) is the name of a village and rural district (*kūra*) between Damascus and Baalbek, see e.g. YāBul 2/913; the reading Zabadān, however, is not certain—the manuscript's ربداى is ambiguous, and there is no parallel passage in the *Paradise of Wisdom* (the German translator of the text also transliterates "Zabadan" [ṬabHT 118], without any explanation).

بخل خمر لتا معتدلا ويعجن بماء مغلى وينزل ساعة حتى يختمر ثم يختضب به بعد غسل الرأس وينشف بمنديل جاف أو يغلى بورق السلق أو غيره وينزل قدر أربع ساعات من النهار ثم يغسل فإنه يصير أسود حالكا ويتبع السواد، إن أصاب الجلد بدقيق الحمص المبلول فإنه يقلعه.

§254 صفة الحب البيمارستاني النافع من القولنج ويخرج الماء الأصفر وينفع الأصحاء ويخرج الفضول، يؤخذ صبر وإهليلج أصفر وسكبينج وتربد مدقوق من كل واحد ثلاثة دراهم عنزروت نصف درهم يسحق وينخل ويعجن بعسل منزوع الرغوة، الشربة مثقالان بماء فاتر.

§255 دواء للرمد جيد مجرب، يؤخذ ماميثا وعنزروت وزعفران من كل واحد جزء ويدق وينخل بحريرة ويذر به.

§256 لعوق جيد نافع للمعدة يقويها ويسخن البدن وينشف البلل، يؤخذ من العسل رطلان ومن ماء السفرجل رطلان ومن خل خمر عتيق رطل ويطبخ الجميع على جمر حتى يصير كالعسل ثم ينزع رغوته ويلقى عليه من الزنجبيل المسحوق أربع أواق ومن الفلفل أوقيتان ويحرك بخشبة فإذا اختلط الجميع أنزل عن النار وسقيت منه.

§257 دواء نافع للصرع والأخلاط الخبيثة والرياح والجنون مجرب فائق، يؤخذ من الحرمل أربعة أجزاء ومن الجاوشير جزء يدق ويجمع بالميبختج وهو الرب ويصير في جام زجاج ويوضع في الشمس

---

٢ أو[1] : و Ox ٥ تربد : بربد Ox || ثلاثة دراهم : ثلثلم Ox ١٤ بالميبختج : بالمسحح Ox

---

128 Whilst the (generic) name "*hospital* pill" (*ḥabb bīmāristānī*) is not as such recorded elsewhere in early Arabic medico-pharmaceutical literature, there is no reason to question the authenticity, currency and provenance of this drug. In Ṭabarī's days, in Iraq, there existed only one hospital, which had been founded in the late 790s CE by the Barmakids, under the aegis of Hārūn ar-Rašīd; it was situated in the southwestern quarter of Baghdad, originally called *Barāmika* hospital (later simply *old* hospital), and seems to have remained the only institution of its kind for at least two generations, possibly even for a century, see e.g. DuBīm 1223a, SSMed 934 and BlaBB 76—it is there that the above drug seems to have been 'invented', or at least thought to have originated. Another possibility, though in my opinion less likely, is the hospital of Gondēšāpūr, whose director (*ra'īs*) was personally known to Ṭabarī, see ṬabFir 39,10.

wine vinegar so as to form an even paste, knead that with boiled water, and put it aside for a while, until it ferments; then, after having washed the head and rubbed it with a dry towel, apply (the paste) as a dye. Alternatively, one can wash (the head) with something like beet leaves, once they have been boiled and then put aside for about four day(light) hours—(these leaves) will turn pitch-black, and the blackness (of the hair) follows from that. Wetted (black) chickpea meal (however), when brought in contact with the skin, detaches it.

Instruction (on how to make) the *hospital* pill[128] which is useful against colic, expels yellow water,[129] works (also) to the advantage of healthy people, and dislodges residues: take three *dirham* each of aloe, yellow myrobalans, sagapenum and ground turpeth (root), and half a *dirham* of sarcocolla, pound that, strain it, and knead it with combless honey; the dose is two *miṯqāl*, with tepid water. § 254

A good, well-tried remedy for (the treatment of) eye inflammation: take one part each of horn poppy, sarcocolla and saffron, grind that, strain it through a cloth of silk, and sprinkle it (on the eye). § 255

A good linctus which serves to strengthen the stomach, to warm the body, and to dry up fluids: take two *raṭl* of honey, two *raṭl* of quince water and one *raṭl* of aged wine vinegar, and cook all that on embers until it gains a honey-like (viscosity); then remove the froth, throw into (the concoction) four *ūqīya* of pounded ginger and two *ūqīya* of pepper, and stir it with a stick; when everything is well mixed, take it off the fire, and let (the patient) drink from it. § 256

An excellent, well-tried remedy which is useful to (treat) epilepsy, vicious (humoral) mixtures, winds,[130] and madness: take four parts of Syrian rue and one part of opopanax, grind that, unite it with the syrup (called) *maibuḫtaǧ*,[131] put (the mixture) into a vessel of glass, and (the vessel) into § 257

---

129 *māʾ aṣfar* "yellow water" refers to ascitic fluid.

130 *riyāḥ* "winds", in the context of epilepsy and madness, denotes an affliction of the nervous system; on this essentially Ayurvedic concept cf. note 108 above.

131 *maibuḫtaǧ* < Persian *may-puḫtah* "inspissated grape wine" (ṬabHT 120 "Weintraubengelee"), see e.g. SchṬab 492 no. 749.

وكلما كاد أن يجف زدتَه من الرب تفعل ذلك به ثلاث مرات ثم يستعمل وكل ما عتق كان أجود، والشربة منه قدر حمصة إلى باقلاة على قدر قوة الشباب وللصبيان قدر عدسة.

§258 دواء نافع لمن قد شبكته الرياح، يؤخذ من ماء الحندقوقا الذي قد برز من غير أن يصب عليه ماء رطل ويجعل في طنجير ويصب عليه مثله من الزيت الركابي ويوقد تحته ويساط حتى يذهب ماء الحندقوقا ويبقى الزيت وينزل ويبرد ويصير في دبة، ويسقى منه الشيخ في كل يوم زنة درهمين إلى ثلاثة دراهم بقدح من نبيذ أو بماء حمص ولا يأكل شيئا إلى نصف النهار ثم يأكل زيرباجة أو تربدة أو ماء حمص ويجتنب المالح والبقول والحموضات ويسقى منه الصبي والمرأة زنة درهمين، وإن أخذ الحندقوقا فطبخ بماء وعصر من مائه على لدغ الهوام وعضة الكلْب الكلِب نفعها وإن شَرِبَه أيضا نفعه.

§259 دواء نافع من عضة الكلْب الكلِب، ينبغي أن يوسع موضع العضة ولا تترك أن تلتحم ويوضع عليها الثوم والملح أو الثوم وسمن البقر العتيق مسخنا ثم يعالج بهذا الدواء أخلاطه: يؤخذ الذراريح تقطع رؤسهن وأرجلهن وأجنحتهن وتنقع في الدوغ وتجفف في الظل وتدق وتنخل بخرقة صفيقة ويؤخذ منها جزء ومن العدس المقشر جزآن ويلت ذلك بشراب جيد قوي ويتخذ منه قرص كل قرص وزنه دانقان، الشربة منه قرص بماء فاتر أو بشراب وإذا شَرِبَ استقبل عين الشمس بعد طلوعها بساعة ويلبس ثيابا كثيرة ويمشي مشيا سريعا متعبا حتى يعرق، فإن وجد غما شرب سكرجة من دهن الزيت السخن أو سمن البقر المسخن، فإن احتبس البول استنقع في ماء حار، وعلامة الشفاء أن يبول الدم، ويكون طعامه مرقة دسمة بلا لحم ولا يأكل أحد من فضلة طعامه ولا يشرب من فضلة شرابه، ولا يعالج موضع العضة إلا بعد خمسة عشر يوما ويقوّر حولها لئلا تلتحم، وإن

---

٢ أجود : + له Ox ٣ ماء[2] : + منه يوخذ Ox ٤ يصب : يضب Ox ٦ دراهم ثلاثة : ثلثلم Ox ٧ تربدة أو : بربده او Ox, تزيده ṬabFir 493,5 || يجتنب : يجنب Ox ٨ لدغ : لذع Ox || نفعها : نفعه Ox ١٠ عليها : عليه Ox ١١ تقطع : يقطع Ox ١٢ تنقع : ينقع Ox ١٥ وجد غما : قد عمل ṬabFir 447,5 : وحده Ox ١٨ تلتحم : يلتحم Ox

١٠ تترك أن تلتحم : ينزل حتى يلتحم ṬabFir 446,18 f. : || Ox

the sun; whenever (the mixture) is almost dry, you add some (more) of the syrup, and you do this three times; then it can be used, although it gets better with age; the dose lies between the size of a chickpea and that of a broad bean, depending on how strong the youngsters are; for children, (the dose is) lentil-sized.

A useful remedy for those who are pestered by (intestinal) winds: take one *raṭl* from the water of sweet clover—the kind that germinated without having been irrigated—, put it in a cauldron, pour onto it the same (amount) of Syrian olive oil,[132] light a fire underneath, and beat (the mixture) until the clover water disappears and (only) the olive oil remains; (then) take it off (the fire), let it cool down, and transfer it into an oil jar. An old man drinks that in a dose of two to three *dirham* daily, with a cupful of fermented juice or chickpea water; he must not eat anything until midday, when he can have spoon-meat or one (mashed) turpeth (root) or (more) chickpea water; and he should avoid all salty or sour (food), as well as greenstuffs. A child or a woman drink it in a dose of two *dirham*. If one takes sweet clover, cooks it in water and squeezes its juice over vermin stings or over the bite of a rabid dog, it is useful; and if (the victim) drinks that (juice), it helps him, too. §258

A remedy which is useful against the bite of a rabid dog. (First), it is necessary to widen the area of the bite, and not leave it to cicatrize—put upon it garlic and salt, or garlic and old, warmed-up ghee. Then treat (the patient) with a drug whose mixture is (as follows): take cantharides, cut off their heads, legs and wings, soak them in whey, dry them in the shade, grind them, strain them through a thick piece of cloth, (then) take from them (a share of) one part and from peeled lentils (a share of) two parts, moisten that with good, strong wine, and turn (this mixture) into pastilles weighing two *dāniq* each; the dose is one pastille, with tepid water or with wine. (The patient), when ingesting (the drug), should face the eye of the sun, one hour after its rise; he should wear a lot of clothes, and go on fast, tiring walks so as to break sweat; if he feels sad, he should swallow a bowlful of warm olive oil or warmed-up ghee; if (he suffers from) urinary retention, he should immerse himself in warm water—(for) a sign of recovery is urinating blood; and his diet should consist of fatty, meatless broths, but nobody must eat from his leftover food, nor drink from his leftover drink! The bite wound (itself) is not treated until after fifteen days—(initially), one (only) excises the surround- §259

132 *zait rikābī* is olive oil that is brought on camels (*rikāb*) from Syria, see LaLex 3/1144b.

أخرج الدم ساعة العض فهو أفضل، ويسقى الماء بأنبوبة سِرًّا مغطّى لأنه لا يقدر أن ينظر إلى الماء، ومنهم من يرى في الماء مَثَل الكلاب ويبول حيوانيا في خلقة جَرْى الكلاب.

§260 صفة حب المسك الهندي وهو دواء تستعمله الملوك مجرب جيد نافع، يؤخذ من الأرمال ومن الحبر من كل واحد رطل يرضان ويغسلان ويصيران في قدر نظيفة ويصب عليهما من الماء العذب أربعين رطلا ويطبخان بنار لينة حتى يبقى من الماء قدر خمسة أرطال ثم ينزل عن النار ويصفى ثم يرد إلى قدر نظيفة مطينة بطين حر وشعر من خارجها أو ما يقوم مقام الشعر ويوقد تحتها نار لينة ويحرك لئلا يحترق حتى يصير بمنزلة العسل الخاثر ثم ينزل عن النار ويجفف في الظل بمنزلة الصبر المغسول فإذا أردتَ أن تتخذ منه الحب أخذتَ منه عشرين مثقالا فتدقه وتنخله بحريرة وتأخذ من القاقلة الصغيرة وهى الهال ومن القرنفل والجوزبوا والبسباسة والعود الصرف الهندي والساذج الهندي والقيطفور وهو دواء هندي يشبه الترمس ومن الصندل الأصفر والهرنوة والكبابة من كل واحد مثقال ومن المسك خمسة مثاقيل ومن الكافور عشر مثاقيل يدق كل واحد على حدته ثم يخلط ويؤخذ من الأرمال خمس مثاقيل فيرض ويغلى بنصف رطل ماء عذب حتى يبقى من الماء أوقيتان ويعجن به الأدوية ويحبب أمثال الحمص ويستعمل إن شاء الله تعالى، والأرمال والحبر من أدوية الهند، وهذا الحب يشد اللثة ويقوي المعدة ويطيب النكهة ويذهب الخمار ورائحة النبيذ ويهضم الطعام.

---

٢ حيوانيا : حيوانا Ox || جرى : حرا Ox ٧ يحرك لئلا يحترق حتى يصير : تحرك ليلا تحترق حتي تصير Ox ٩ القاقلة الصغيرة وهى : الفافله الصغار وهو Ox ١٤ الحبر : الخبر Ox

---

133 The following compound preparation is not recorded elsewhere in Arabic medico-pharmaceutical literature (cf. also note 137 below), and therefore it seems that Ṭabarī himself baptized the drug, using, as was customary, the name of a key ingredient (musk), and adding the name of its place of origin (India). On musk in early Arabic pharmacy now see KiGP 303–317.

134 *qāqulla* (*ṣaġīra*) is a synonym of *hāl* (< Sanskrit *elā* < [?] Tamil *ēlam*), see SchṬab 335 ff. no. 561 and 518 no. 781.

135 *qytfwr* (قيطفور sic expressim) is a Sanskrit term I have not been able to identify.

136 Cf. note 13 above.

ing area, to prevent (the wound) from cicatrizing; it is best to force blood out (of the wound) straight after the bite. (The patient) should be made to drink through a reed tube (from a bowl that is) covered and concealed, because he cannot bear the sight of water. There are some (victims) who see dog-like shapes in the water, and (others who) piss in the manner of an animal, just as dogs normally do.

Instruction (on how to make) the *Indian musk* pill[133] which is a good, efficient, well-tried remedy employed by kings: take one *raṭl* each of sand and (solid) ink, crush them both, wash them, transfer them into a clean pot, pour forty *raṭl* of fresh water over them, and cook them on a low flame until about five *raṭl* of water are left; then lift (the pot) off the fire, and pass (the liquid) through a filter; then return it to (another) clean pot whose outside is coated with (a mixture of) pure clay and hair—or something akin to hair—, light a gentle fire underneath it, stir (the liquid) to prevent it from burning, (and do that) until it becomes like thickened honey; then lift it off the fire, and put it in the shade to dry, (until it becomes) like the washed (pulp of) aloe (leaves). Now, when you want to make the pills, take twenty *miṯqāl* from (the congealed product), grind that, and strain it through a cloth of silk; (further) take one *miṯqāl* each of *qāqulla ṣaġīra*—which is cardamom[134]—, clove, nutmeg, mace, pure Indian lignaloes, Indian cassia, *qytfwr*—an Indian drug resembling the lupine[135]—, yellow sandalwood, aloe berries and cubeb, as well as five *miṯqāl* of musk and ten *miṯqāl* of camphor; grind each (ingredient) separately before mixing (everything) together; (then) take five *miṯqāl* of sand, crush it, and boil it in half a *raṭl* of fresh water until two *ūqīya* of that water are left; (then) knead the (above) ingredients with (the sand-water), roll (that mixture) into pills the size of chickpeas, and use them—God the Sublime willing.[136] Sand and ink are among the medicinal substances (used) by the Indians.[137] The (aforesaid) pills tighten the gums, strengthen the stomach, improve the smell of the breath, get rid of a hangover, (conceal) the stench of wine, and promote the digestion of food. § 260

137 This ostensibly Indian prescription, featuring sand (*armāl* [sic]) and ink (*ḥibr*) as characteristic ingredients, cannot be substantiated in Ṭabarī's Ayurvedic sources; the very precision of its pharmacological instructions seems to suggest a piece of oral information. The fabrication of this compound 'pill' (*ḥabb*) is not attested in the *Paradise of Wisdom*. It may be remarked here in passing that both sand (ἄμμος) and ink (μέλαν) are registered as therapeutic agents, for example, by Dioscorides (DioWe 3/102 no. 148 and 108 no. 162), though obviously not in the framework of a *pharmacological* prescription.

§261 وهذه صورة هندسية أربعة في أربعة تجمعْ ما في كل صنف منها:

عنبر ستة أجزاء	كافور خمسة أجزاء	سك ثلاثة أجزاء	مسك عشرة أجزاء	دخنة
عود ستة عشر جزءً	صندل جزآن	زعفران وفلنجة ثلاثة عشر جزءً نصفين	سنبل ستة أجزاء	دخنة
جوزبوا ستة أجزاء كافور خمسة أجزاء	مسك سبعة أجزاء	تغر طيب هندي ستة عشر جزءً	قرفة سبعة أجزاء قرنفل سبعة أجزاء	ذريرة
بسباسة ستة أجزاء	قاقلة ستة أجزاء هرنوة ستة أجزاء	قرفة وقرنفل ثلاثة أجزاء نصفين	كبابة يمانية ثمانية أجزاء قسط مثله	طلْى الحمام
طلْى الحمام	حب يؤخذ في الفم	ذريرة	طلْى الحمام	

١ هندسية : هندسة Ox ‖ منها : منه Ox ٢ سك : سكر Ox, ṬabFir 602 ٥ فلنجة : فلـحه Ox ٨ تغر : تعز Ox, نعـــــسم ṬabFir 602 (cf. ibid. 610 بعر [!] ter) ‖ قرفة : قرفا Ox ١٢ قرفة : قرفا Ox

138 *sukk* is a pastille-shaped aromatic compound, prepared from a dry, dark-coloured mixture of gallnuts and emblics, called *rāma/ik*, and often (but not necessarily) involving musk, see NaTTB 565; for an early mention of *sukk* see KinAq 294 no. 171; further LaLex 3/1158c–1159a and 4/1387b. The etymologies of both *sukk* and *rāma/ik* are uncertain—the former term may go back to Sanskrit *śuka* "a partic[ular] drug and perfume" (MWDic 1079c), or perhaps to *sukha* "a drug or medicinal root called *vṛiddhi*" (MWDic

And here is a geometrical diagram, four by four—add up what is in each of its boxes: § 261

incense	musk ten parts	*sukk*[138] three parts	camphor five parts	ambergris six parts
incense	spikenard six parts	saffron and *falanǧa*[139] thirteen parts in two (equal) halves	sandalwood two parts	lignaloes sixteen parts
powder	canella seven parts clove seven parts	*taġar*[140] —an Indian perfume— sixteen parts	musk seven parts	nutmeg six parts camphor five parts
bath lotion	Yemenite cubeb eight parts costmary the same	canella and clove three parts in two (equal) halves	cardamom six parts aloe berries six parts	mace six parts
	bath lotion	powder	pill held in mouth	bath lotion

1221a), that is a species of fragrant orchids (*Habenaria intermedia*, see PDIP s.v. *vṛddhi*). A parallel table in the *Paradise of Wisdom* (ṬabFir 602) reads *sukkar* "sugar" instead of *sukk*, and the German translator of the present text has adopted this inferior reading (ṬabHT 126 "Zucker").

139 *falanǧa* cannot be identified with certainty—see the discussion SchṬab 323 f. no. 544, who summarizes a bewildering range of possibilities, insinuating (but not pursuing) an 'Indian' etymology; further BellSA 57 f. (polysemous). The term may in fact go back to Sanskrit *priyaṅgu* "a medicinal plant and perfume [...] described in some places as a fragrant seed", see MWDic 711a with PDIP s.v.; the identification *φαλάγγιον* "spider-wort", suggested FaddNC 345 no. 5, is certainly wrong.

140 *taġar*, described here as "an Indian perfume" (*ṭīb hindī*), is an emendation of the manuscript's تعز (a parallel table in the *Paradise of Wisdom* [ṬabFir 602] reads ٮعسم, and on three other occasions [ibid. 610], also forming part of a tabular registry, the term comes up as بعر "animal dung", which hardly fits in with the given contexts); the German translator of the present text says that the term is indecipherable ("nicht zu entziffern"), but refers in a note (ṬabHT 127) to SchṬab 129 f. no. 168 where the correct etymology is already suggested: *taġar* < Sanskrit *tagara*[*ka*], which is identified MWDic 431 f. as "*Tabernaemontana coronaria* [scil. pinwheel flower] and a fragrant powder prepared from it" and, perhaps more accurately, PDIP s.v. *tagara* "*Valeriana jatamansi* [scil. Indian valerian]".

Bibliography

This bibliography contains *cited* literature only.

AriDA *Aristotle. De Anima*, ed., tr. and comm. R.D. Hicks, Cambridge (Cambridge University Press) 1907.

ArnDA Rüdiger Arnzen, *Aristoteles' De Anima. Eine verlorene spätantike Paraphrase in arabischer und persischer Überlieferung*, Leiden–NewYork–Köln (Brill) 1998 (Aristoteles Semitico-Latinus 9).

AzKM ʿAbdallāh ibn Muḥammad al-Azdī, *Kitāb al-Māʾ*, 1–3, ed. Hādī Ḥasan Ḥammūdī, [Masqaṭ] (Wizārat at-Turāṯ al-Qaumī waṯ-Ṯaqāfa) 1416/1996.

BadUṣ *al-Uṣūl al-yūnānīya lin-naẓarīyāt as-siyāsīya fī l-islām* [part 1], ed. ʿAbdarraḥmān Badawī, al-Qāhira (Maktabat an-Nahḍa al-Miṣrīya) 1373/1954 (Dirāsāt Islāmīya 15).

BaiTat ʿAlī ibn Abī l-Qāsim al-Baihaqī, *Taʾrīḫ al-ḥukamāʾ al-musammā bi-Durrat al-aḫbār wa-lumʿat al-anwār yaʿnī tarǧuma-ye Tatimma-ye Ṣiwān al-ḥikma*, ed. Muḥammad Šafīʿ, Lāhūr 1350[/1931].

BeDio J[ulius] Berendes, *Des Pedanios Dioskurides aus Anazarbos Arzneimittellehre in fünf Büchern*, Stuttgart (Ferdinand Enke) 1902.

BeInd *AlBeruni's India. An Account of the Religion, Philosophy, Literature, Geography, Chronology, Astronomy, Customs, Laws and Astrology of India about A.D. 1030*, 1–2, tr. Edward C. Sachau, London (Kegan Paul, Trench, Trübner & Co.) 1910 (Trübner's Oriental Series).

BellSA Jamal Bellakhdar, "La *falanǧah*, un simple aromatique de la pharmacopée arabo-islamique de nature non identifiée: essai de détermination", in: *al-Andalus–Magreb* 22 (2015), pp. 29–63.

BergḤu *Ḥunain ibn Isḥāq über die syrischen und arabischen Galen-Übersetzungen*, ed. and tr. G[otthelf] Bergsträsser, Nendeln/Liechtenstein (Kraus) 1966 (reprint of the edition Leipzig 1925) (Abhandlungen für die Kunde des Morgenlandes 17,2).

BīFih Muḥammad ibn Aḥmad al-Bīrūnī, *Risāla fī Fihrist kutub Muḥammad ibn Zakarīyāʾ ar-Rāzī*, in: *Fehrest-e ketābhā-ye Rāzī ve-nāmhā-ye ketābhā-ye Bērūnī*, ed. and tr. [into Persian] Mehdī Moḥaqqeq, Tehrān (Moʾassasa-ye Entešārāt ve-Čāp Dānešgāh-e Tehrān) 1366[š/1987] (Entešārāt-e Dānešgāh-e Tehrān 1406), pp. 1–42.

BlaBB Kevin van Bladel, "The Bactrian Background of the Barmakids", in: *Islam and Tibet. Interactions along the Musk Routes*, ed. Anna Akasoy, Charles Burnett and Ronit Yoeli-Tlalim, Farnham/Surrey (Ashgate) 2011, pp. 43–88.

BosṬab C.E. Bosworth, "al-Ṭabarī", in: EI² [q.v.] 10/11a–15b.

BouTP Véronique Boudon-Millot, *Galien: Thériaque à Pison. Texte établi et traduit*, Paris (Les Belles Lettres) 2016 (Collection des Universités de France 526).

BroFḤ C[arl] Brockelmann, review of "Siddiqi (ed.) *Firdausu 'l-Ḥikmat*", in: *Zeitschrift für Semitistik und Verwandte Gebiete* 8 (1932), pp. 270–288.

BrowAM Edward G. Browne, *Arabian Medicine. Being the Fitzpatrick Lectures Delivered at the College of Physicians in November 1919 and November 1920*, Cambridge (Cambridge University Press) 1921.

BryKḤ Jennifer S. Bryson, *The Kitāb al-Ḥāwī of Rāzī (ca. 900 AD): Book One of the Ḥāwī on Brain, Nerve, and Mental Disorders. Studies in the Transmission of Medical Texts from Greek into Arabic into Latin*, Ann Arbor (Bell & Howell) 2001.

BuḫAD *'Ubaidallāh Ibn Buḫtīšū' on Apparent Death*, ed. and tr. Oliver Kahl, *with a Hebrew supplement* by Gerrit Bos, Leiden–Boston (Brill) 2018 (Islamic Philosophy, Theology and Science 105).

CaSaṃ *Caraka Saṃhitā (with a critical exposition based on Cakrapāṇidatta's Āyurvedadīpikā)*, 1–7, ed. and tr. Ram Karan Sharma and Vaidya Bhagwan Dash, Varanasi (Chowkhamba Sanskrit Series Office) 2009–2011 (reprint) (Chowkhamba Sanskrit Studies 94).

CRCha J.F. McCurdy and R.W. Rogers, "Chaldea", in: *JE* [q.v.] 3/661b–662b.

DimRH *La Risāla al-Hārūniyya de Masīḥ b. Ḥakam al-Dimašqī*, ed. and tr. Suzanne Gigandet, Damas (Institut Français d'Études Arabes de Damas) 2001.

DioWe *Pedanii Dioscuridis Anazarbei De Materia Medica Libri Quinque*, 1–3, ed. Max Wellmann, Berolini (Weidmann) 1906–1914.

DīwIH *Dīwān Ibrāhīm Ibn Harma*, ed. Muḥammad Ǧabbār al-Mu'aibid, an-Naǧaf (Maṭba'at al-Ādāb) 1389/1969.

DoSupp R[einhart] Dozy, *Supplément aux dictionnaires arabes*, 1–2, Leyde (E.J. Brill) 1881.

DuBīm D.M. Dunlop, "Bīmāristān", in: EI² [q.v.] 1/1222b–1224b.

EbThPW^{1,2} *The Polemical Works of 'Alī al-Ṭabarī* [scil. *ar-Radd 'alā n-naṣārā* and *ad-Dīn wad-daula*], ed. [and tr.] Rifaat Ebied and David Thomas, Leiden–Boston (Brill) 2016 (The History of Christian-Muslim Relations 27).

EI¹ *Enzyklopaedie des Islam. Geographisches, ethnographisches und biographisches Wörterbuch der muhammedanischen Völker*, 1–5, ed. M.Th. Houtsma et al., Leiden–Leipzig (E.J. Brill & Otto Harrassowitz) 1908–1938.

EI² *The Encyclopaedia of Islam. New Edition*, 1–12, ed. C.E. Bosworth et al., Leiden (Brill) 1960–2004.

EI³ *Encyclopaedia of Islam: Three*, 1–, ed. Kate Fleet et al., Leiden–Boston (Brill) 2007– [online at: referenceworks.brillonline.com/browse/encyclopaedia-of-islam-3].

EIr *Encyclopaedia Iranica*, 1–, ed. Elton Daniel et al., Leiden–Boston (Brill) 1985– [online at: iranicaonline.org].

FaddNC J.-M. Faddegon, "Notice critique sur le *Firdausu 'l-Ḥikmat or Paradise of Wisdom* de 'Alí b. Rabban al-Ṭabarí", in: *Journal Asiatique* 218 (1931), pp. 327–352.

FiCG Gerhard Fichtner et al., *Corpus Galenicum. Bibliographie der galenischen und pseudogalenischen Werke*, Berlin (Berlin-Brandenburgische Akademie der Wissenschaften) 2019 (enlarged and corrected edition) [online at: cmg.bbaw.de/online-publications/Galen-Bibliographie_2019-01.pdf].

FiCH Gerhard Fichtner et al., *Corpus Hippocraticum. Bibliographie der hippokratischen und pseudohippokratischen Werke*, Berlin (Berlin-Brandenburgische Akademie der Wissenschaften) 2017 (enlarged and corrected edition) [online at: cmg.bbaw.de/online-publications/Hippokrates-Bibliographie_2017-12.pdf].

FMDey Wolfgang Felix and Wilferd Madelung, "Deylamites", in: *EIr* [q.v.] 7/342a–347a.

FoPM Adolf Fonahn, *Zur Quellenkunde der persischen Medizin*, Leipzig (Johann Ambrosius Barth) 1910.

GabAd F. Gabrieli, "Adab", in: *EI*² [q.v.] 1/175b–176a.

GalHe *Galeni De Temperamentis Libri III*, ed. Georgius Helmreich, Lipsiae (B.G. Teubner) 1904.

GalKo *Galeni De Sanitate Tuenda*, ed. Konradus Koch, Lipsiae–Berolini (B.G. Teubner) 1923 (Corpus Medicorum Graecorum 5.4,2) [pp. 3–198].

GalKü *Claudii Galeni Opera Omnia*, 1–20, ed. and tr. [into Latin] Carolus Gottlob Kühn, Lipsiae (Car[olus] Cnobloch) 1821–1833 (Medicorum Graecorum Opera Quae Exstant).

GaS Fuat Sezgin, *Geschichte des arabischen Schrifttums*, 1–17, Leiden (E.J. Brill) 1967–1984 [then] Frankfurt am Main (Institut für Geschichte der Arabisch-Islamischen Wissenschaften) 2000–2015.

ĞauḎH 'Abdarraḥmān ibn 'Alī Ibn al-Ğauzī, *Ḏamm al-hawā*, ed. Ḫālid 'Abdallaṭīf as-Sab' al-'Alamī, Bairūt (Dār al-Kitāb al-'Arabī) 1431/2010.

GhFir Antonella Ghersetti, *Il Kitāb Arisṭāṭalīs al-faylasūf fī l-firāsa nella traduzione di Ḥunayn b. Isḥāq*, Roma (Herder) 1999 (Quaderni di Studi Arabi 4).

GroBad Heinz Grotzfeld, *Das Bad im arabisch-islamischen Mittelalter. Eine kulturgeschichtliche Studie*, Wiesbaden (Otto Harrassowitz) 1970.

GuGT Dimitri Gutas, *Greek Thought, Arabic Culture. The Graeco-Arabic Translation Movement in Baghdad and Early 'Abbāsid Society (2nd–4th/8th–10th Centuries)*, London–New York (Routledge) 1998.

GuWL Dimitri Gutas, *Greek Wisdom Literature in Arabic Translation. A Study of the*

Graeco-Arabic Gnomologia, New Haven/Conn. (American Oriental Society) 1975 (American Oriental Series 60).

ḪaWaf Aḥmad ibn Muḥammad Ibn Ḫallikān, *Wafayāt al-aʿyān wa-anbāʾ abnāʾ az-zamān*, vol. 4, ed. Muḥammad Muḥyīddīn ʿAbdalḥamīd, al-Qāhira (Maktabat an-Nahḍa al-Miṣrīya) 1367/1948.

HeBā Ernst Herzfeld, "Bābil", in: *EI*[1] [q.v.] 1/570a–571b.

HiMG Walther Hinz, *Islamische Masse und Gewichte umgerechnet ins metrische System*, Leiden (E.J. Brill) 1955 (Handbuch der Orientalistik, Ergänzungsband 1,1).

HippHe[1] *Hippocratis De Prisca Medicina*, ed. I.L. Heiberg, Lipsiae–Berolini (B.G. Teubner) 1927 (Corpus Medicorum Graecorum 1.1).

HippHe[2] *Hippocratis De Aere Locis Aquis*, ed. I.L. Heiberg, Lipsiae–Berolini (B.G. Teubner) 1927 (Corpus Medicorum Graecorum 1.1).

HippHe[3] *Hippocratis De Alimento*, ed. I.L. Heiberg, Lipsiae–Berolini (B.G. Teubner) 1927 (Corpus Medicorum Graecorum 1.1).

HippJo *Hippocratis De Diaeta*, ed. and tr. Robert Joly, Berlin (Akademie Verlag) 2003 (Corpus Medicorum Graecorum 1.2,4).

HippKü *Hippocratis Opera Quae Feruntur Omnia*, vol. 1, ed. Hugo Kühlewein, Lipsiae (B.G. Teubner) 1894.

HippLi *Œuvres complètes d'Hippocrate*, 1–10, ed. and tr. É[mile] Littré, Amsterdam (Adolf M. Hakkert) 1961–1962 (reprint of the edition Paris 1839–1861).

HoyPo Robert Hoyland, "A New Edition and Translation of the Leiden Polemon", in: *Seeing the Face, Seeing the Soul. Polemon's Physiognomy from Classical Antiquity to Medieval Islam*, ed. Simon Swain, Oxford (Oxford University Press) 2007, pp. 329–463.

IAU *A Literary History of Medicine: The ʿUyūn al-anbāʾ fī ṭabaqāt al-aṭibbāʾ of Ibn Abī Uṣaybiʿah*, 1–3 [in 5 vols], ed. and tr. Emilie Savage-Smith, Simon Swain, Geert Jan van Gelder et al., Leiden–Boston (Brill) 2020 (Handbook of Oriental Studies, Section One, The Near and Middle East 134).

IǦṬab Sulaimān ibn Ḥassān Ibn Ǧulǧul, *Ṭabaqāt al-aṭibbāʾ wal-ḥukamāʾ* (*Les Générations des médecins et des sages*), ed. Fuʾād Sayyid, Le Caire (Institut Français d'Archéologie Orientale) 1955 (Textes et Traductions d'Auteurs Orientaux 10).

InkḤā Bruce Inksetter, "al-Ḥārith b. Kalada", in: *EI*[3] [q.v.] s.n.

ISBio *Ibn Saad. Biographien der mekkanischen Kämpfer Muhammeds in der Schlacht bei Bedr*, vol. 3.1, ed. Eduard Sachau, Leiden (E.J. Brill) 1904.

IsHṬ Edward G. Browne, *An Abridged Translation of the History of Ṭabaristán Compiled about A.H. 613 (A.D. 1216) by Muḥammad b. al-Ḥasan b. Isfandiyár*, Leyden–London (E.J. Brill and Bernard Quaritch) 1905 (E.J.W. Gibb Memorial Series 2).

ISQā al-Ḥusain ibn ʿAbdallāh Ibn Sīnā, *al-Qānūn fī ṭ-ṭibb*, 1–3, Bairūt (Dār Ṣādir) n.d. (reprint of the edition Būlāq [1294/1877]).

JE *The Jewish Encyclopedia. A Descriptive Record of the History, Religion, Literature, and Customs of the Jewish People from the Earliest Times to the Present Day*, 1–12, ed. Isidore Singer et al., NewYork–London (Funk & Wagnalls Company) 1901–1906.

JobBT *Encyclopædia of Philosophical and Natural Sciences as Taught in Baghdad about A.D. 817 or Book of Treasures by Job of Edessa*, ed. and tr. A[lphonse] Mingana, Cambridge (W. Heffer & Sons) 1935 (Woodbrooke Scientific Publications 1).

JoIM Julius Jolly, *Indian Medicine*, tr. from German and suppl. with notes by C.G. Kashikar, New Delhi (Munshiram Manoharlal) 2012 (reprint of the 2nd revised edition 1977).

KaAnt Oliver Kahl, "Two Antidotes from the 'Empiricals' of Ibn at-Tilmīḏ", in: *Journal of Semitic Studies* 55 (2010), pp. 479–496.

KaIMT Oliver Kahl, "On the Transmission of Indian Medical Texts to the Arabs in the Early Middle Ages", in: *Arabica* 66 (2019), pp. 82–97.

KaWM Oliver Kahl, "ʿĪsā ibn Māssa on Medicinal Weights and Measures", in: *Orientalia Lovaniensia Periodica* 23 (1992), pp. 275–279.

KeMut H. Kennedy, "al-Mutawakkil ʿalā 'llāh", in: *EI*^{2} [q.v.] 7/777b–778a.

KiGP Anya H. King, *Scent from the Garden of Paradise. Musk and the Medieval Islamic World*, Leiden–Boston (Brill) 2017 (Islamic History and Civilization 140).

KinAq *The Medical Formulary or Aqrābādhīn of al-Kindī*, ed. [facsimile], tr. and comm. Martin Levey, Madison–Milwaukee–London (The University of Wisconsin Press) 1966 (Publications in Medieval Science).

LaLex Edward William Lane, *An Arabic-English Lexicon*, 1–8, London–Edinburgh (Williams & Norgate) 1863–1893.

LBʿUm G. Levi Della Vida and M. Bonner, "ʿUmar b. al-Khaṭṭāb", in: *EI*^{2} [q.v.] 10/818b–821a.

LeʿA B. Lewis, "ʿAbbāsids", in: *EI*^{2} [q.v.] 1/15a–23b.

LecHis Lucien Leclerc, *Histoire de la médecine arabe. Exposé complet des traductions du grec. Les sciences en Orient, leur transmission à l'Occident par les traductions latines*, 1–2, Paris (Ernest Leroux) 1876.

LeSEC G[uy] Le Strange, *The Lands of the Eastern Caliphate. Mesopotamia, Persia, and Central Asia from the Moslem Conquest to the Time of Timur*, Cambridge (Cambridge University Press) 1905 (Cambridge Geographical Series).

LSLex Henry George Liddell and Robert Scott, *A Greek-English Lexicon*, Oxford (Clarendon Press) 1996 (9th edition with a revised supplement).

MāNid *Mādhava Nidānam (Roga Viniścaya) of Mādhavakara (A Treatise on Āyur-*

veda), ed. and tr. K.R. Srikantha Murthy, Varanasi (Chaukhambha Orientalia) 2011 (reprint) (Jaikrishnadas Ayurveda Series 69).

MasPO ʿAlī ibn al-Ḥusain al-Masʿūdī, *Les prairies d'or* [*Murūǧ aḏ-ḏahab*], vol. 8, ed. and tr. C. Barbier de Meynard, Paris (Société Asiatique) 1874 (Collection d'Ouvrages Orientaux).

MenHL Everett Mendelsohn, *Heat and Life. The Development of the Theory of Animal Heat*, Cambridge/Mass. (Harvard University Press) 1964.

MeyPW Max Meyerhof, "ʿAlî aṭ-Ṭabarî's 'Paradise of Wisdom', one of the Oldest Arabic Compendiums of Medicine", in: *Isis* 16 (1931), pp. 6–54.

MeyṬab Max Meyerhof, "ʿAlī ibn Rabban aṭ-Ṭabarī, ein persischer Arzt des 9. Jahrhunderts n.Chr.", in: *Zeitschrift der Deutschen Morgenländischen Gesellschaft* 85 (1931), pp. 38–68.

MeyTT *The Book of the Ten Treatises on the Eye Ascribed to Hunain ibn Is-hâq (809–877 A.D.). The Earliest Existing Systematic Text-Book of Ophthalmology*, ed. and tr. Max Meyerhof, Cairo (Government Press) 1928.

MWDic Monier Monier-Williams, *A Sanskrit-English Dictionary*, Oxford (Clarendon Press) 1899.

NaACK Nawal Nasrallah, *Annals of the Caliphs' Kitchens. Ibn Sayyār al-Warrāq's Tenth-Century Baghdadi Cookbook*, Leiden–Boston (Brill) 2010.

NadFih Muḥammad ibn Isḥāq al-Warrāq an-Nadīm, *al-Fihrist*, 1–2, ed. Gustav Flügel [then] Johannes Roediger and August Mueller, Bairūt (Maktabat Ḫaiyāṭ) n.d. (reprint of the edition Leipzig 1871–1872) (Rawāʾiʿ at-Turāṯ al-ʿArabī).

NaTTB Nawal Nasrallah, *Treasure Trove of Benefits and Variety at the Table: A Fourteenth-Century Egyptian Cookbook*, Leiden–Boston (Brill) 2018 (Islamic History and Civilization 148).

OlDDS Joshua Olsson, *Design, Determinism and Salvation in the Firdaws al-Ḥikma of ʿAlī Ibn Rabban al-Ṭabarī*, PhD-thesis, University of Cambridge/UK 2015 [online at: academia.edu/28390060].

PDIP *Pandanus Database of Indian Plants*, Prague (Seminar of Indian Studies, Charles University) 1998–2009 [online at: iu.ff.cuni.cz/pandanus/database].

PinAM David Pingree, *The Thousands of Abū Maʿshar*, London (The Warburg Institute) 1968 (Studies of the Warburg Institute 30).

PinHay D[avid] Pingree, "ʿIlm al-hayʾa", in: *EI*^{2} [q.v.] 3/1135a–1138a.

PorMel Peter E. Pormann, "Introduction", in: *Rufus of Ephesus. On Melancholy*, ed. Peter E. Pormann, Tübingen (Mohr Siebeck) 2008 (Sapere 12), pp. 3–23.

PSThes R[obert] Payne Smith, *Thesaurus Syriacus*, 1–2, Oxonii (Clarendon Press) 1879–1901.

QalAq Muḥammad ibn Bahrām al-Qalānisī, *Aqrabāḏīn al-Qalānisī*, ed. M.Z. al-

Bābā, Ḥalab (Maʿhad at-Turāṯ al-ʿIlmī al-ʿArabī) 1403/1983 (Maṣādir wa-Dirāsāt fī Taʾrīḫ aṭ-Ṭibb al-ʿArabī 3).

QazKos Zakarīyāʾ ibn Muḥammad al-Qazwīnī, *Kosmographie* [*ʿAǧāʾib al-maḫlūqāt*], vol. 1, ed. Ferdinand Wüstenfeld, Göttingen (Dieterich'sche Verlagsbuchhandlung) 1849.

RaggAP Lucia Raggetti, *ʿĪsā ibn ʿAlī's Book on the Useful Properties of Animal Parts*, Berlin–Boston (De Gruyter) 2018 (Science, Technology, and Medicine in Ancient Cultures 6).

RhaCB Oliver Kahl, *The Sanskrit, Syriac and Persian Sources in the Comprehensive Book of Rhazes*, Leiden–Boston (Brill) 2015 (Islamic Philosophy, Theology and Science 93).

RhaMan Muḥammad ibn Zakarīyāʾ ar-Rāzī, *Manāfiʿ al-aġḏiya wa-dafʿ maḍārrihā*, Miṣr al-maḥmīya [Cairo] (Maṭbaʿa Ḫairīya) 1305[/1888].

RiWaSB Helmut Ritter and Richard Walzer, *Arabische Übersetzungen griechischer Ärzte in Stambuler Bibliotheken*, Berlin (Verlag der Akademie der Wissenschaften, in Kommission bei Walter de Gruyter & Co.) 1934 (Sonderausgabe aus den Sitzungsberichten der Preussischen Akademie der Wissenschaften, Phil[ologisch]-Hist[orische] Klasse 26).

SābAq *Sābūr ibn Sahl. Dispensatorium parvum (al-Aqrābādhīn al-ṣaghīr)*, ed. Oliver Kahl, Leiden–NewYork–Köln (E.J. Brill) 1994 (Islamic Philosophy, Theology and Science 16).

SābDis *Sābūr ibn Sahl. The Small Dispensatory*, tr. Oliver Kahl, Leiden–Boston (Brill) 2003 (Islamic Philosophy, Theology and Science 53).

SaRép Khalil Samir, "La réponse d'al-Ṣafī ibn al-ʿAssāl à la réfutation des chrétiens de ʿAlī al-Ṭabarī", in: *Parole de l'Orient* 11 (1983), pp. 281–328.

SchMP [M.] Schachter, "Un médecin perse du IXe siècle, d'origine chrétienne, Ali ibn Rabban at[-]Tabari", in: *Bulletin de la Société Française d'Histoire de la Médecine* 26 (1932), pp. 165–170.

SchṬab Werner Schmucker, *Die pflanzliche und mineralische Materia Medica im Firdaus al-Ḥikma des Ṭabarī*, Bonn (Selbstverlag des Orientalischen Seminars) 1969 (Bonner Orientalistische Studien 18).

ScPhy *Scriptores Physiognomonici Graeci et Latini*, vol. 1: *Physiognomonica Pseudaristotelis, Graece et Latine, Adamantii cum Epitomis Graece, Polemonis e Recensione Georgii Hoffmanni Arabice et Latine Continens*, ed. Richardus Foerster, Lipsiae (B.G. Teubner) 1893.

SidAA M.Z. Siddiqi, "An Early Arabian Author on the Indian System of Medicine", in: *The Calcutta Review* 41 (1931), pp. 277–283.

SidML M.Z. Siddíqí, *Studies in Arabic and Persian Medical Literature*, Calcutta (Calcutta University Press) 1959.

SiGEF Alfred Siggel, "Gynäkologie, Embryologie und Frauenhygiene aus dem

'Paradies der Weisheit über die Medizin' des Abū Ḥasan ʿAlī b. Sahl Rabban aṭ-Ṭabarī", in: *Quellen und Studien zur Geschichte der Naturwissenschaften und der Medizin* 8 (1941–1942), pp. 216–272.

SiIB Alfred Siggel, *Die indischen Bücher aus dem Paradies der Weisheit über die Medizin des ʿAlī ibn Sahl Rabban aṭ-Ṭabarī*, Wiesbaden (Verlag der Akademie der Wissenschaften und der Literatur in Mainz, in Kommission bei Franz Steiner) 1950 (Abhandlungen der Geistes- und Sozialwissenschaftlichen Klasse 14).

SiPK Alfred Siggel, *Die propädeutischen Kapitel aus dem Paradies der Weisheit über die Medizin des ʿAlī b. Sahl Rabban aṭ-Ṭabarī*, Wiesbaden (Verlag der Akademie der Wissenschaften und der Literatur in Mainz, in Kommission bei Franz Steiner) 1953 (Abhandlungen der Geistes- und Sozialwissenschaftlichen Klasse 8).

SpiWG Otto Spies, review of "Raslan (tr.) *Über die Wahrung* [sic] *der Gesundheit*", in: *Die Welt des Islams* 17 (1976), pp. 255–256.

SpRis Fabrizio Speziale, "La *Risāla al-ḏahabīyya* [sic]. Traité médical attribué à l'imām ʿAlī al-Riżā", in: *Luqmān* 20 (2004), pp. 7–34.

SSMed Emilie Savage-Smith, "Medicine", in: *Encyclopedia of the History of Arabic Science*, 1–3, ed. Roshdi Rashed in collaboration with Régis Morelon, London–NewYork (Routledge) 1996, 3/903–962.

SSNC Emilie Savage-Smith, *A New Catalogue of Arabic Manuscripts in the Bodleian Library, University of Oxford*, vol. 1: *Medicine*, Oxford (Oxford University Press) 2011.

StALJ Moritz Steinschneider, *Die arabische Literatur der Juden. Ein Beitrag zur Literaturgeschichte der Araber, grossenteils aus handschriftlichen Quellen*, Frankfurt a.M. (J. Kauffmann) 1902 (Bibliotheca Arabico-Judaica).

StSSA Moritz Steinschneider, "Sahl ben Bischr, Sahl al-Tabari und Ali b. Sahl", in: *Zeitschrift der Deutschen Morgenländischen Gesellschaft* 54 (1900), pp. 39–48.

SuSaṃ *Illustrated Suśruta Saṁhitā*, 1–3, ed. and tr. K.R. Srikantha Murthy, Varanasi (Chaukhambha Orientalia) 2005–2008 (reprint) (Jaikrishnadas Ayurveda Series 102).

SutMA Heinrich Suter, *Die Mathematiker und Astronomen der Araber und ihre Werke*, Leipzig (B.G. Teubner) 1900 (Abhandlungen zur Geschichte der Mathematischen Wissenschaften mit Einschluss ihrer Anwendungen 10).

ṬabDD ʿAlī ibn Sahl Rabban aṭ-Ṭabarī, *The Book of Religion and Empire (Kitāb ad-Dīn wad-daula). A Semi-Official Defence and Exposition of Islam Written by Order at the Court and with the Assistance of the Caliph Mutawakkil (A.D. 847–861)*, 1–2, ed. and tr. A[lphonse] Mingana, Manchester–London

(Manchester University Press with Longmans, Green & Co. et al.) 1922–1923.

ṬabFir 'Alī ibn Sahl Rabban aṭ-Ṭabarī, *Firdaus al-ḥikma fī ṭ-ṭibb*, ed. Muḥammad Zubair aṣ-Ṣiddīqī, Berlin (Sonnendruckerei) 1928.

ṬabFir[2] 'Alī ibn Sahl Rabban aṭ-Ṭabarī, *Firdaus al-ḥikma fī ṭ-ṭibb*, ed. 'Abdalkarīm Sāmī al-Ǧundī, Bairūt (Dār al-Kutub al-'Ilmīya) 1423/2002.

ṬabHT *Über die Erhaltung der Gesundheit. Ein Hygiene Traktat von Ali ibn Sahl Rabban at-Tabari*, tr. [into German] Usama Raslan, Bonn (Selbstverlag) 1975.

ṬabRa I.-A. Khalifé and W. Kutsch, "Ar-Radd 'ala-n-Naṣārā de 'Alī aṭ-Ṭabarī", in: *Mélanges de l'Université Saint Joseph* 36 (1959), pp. 113–148.

ṬabRi Jean-Marie Gaudeul, *Riposte aux chrétiens par 'Alī al-Ṭabarī*, Roma (Pontificio Istituto di Studi Arabi e d'Islamistica) 1995 (Studi Arabo-Islamici 7).

Ṭab[2]Ann Muḥammad ibn Ǧarīr aṭ-Ṭabarī, *Annales* [*Ta'rīḫ ar-rusul wal-mulūk*], vol. 3.2, ed. M.J. de Goeje, Lugd[uni] Bat[avorum] (E.J. Brill) 1881.

ThErk Hans-Jürgen Thies, *Erkrankungen des Gehirns insbesondere Kopfschmerzen in der Arabischen Medizin*, Bonn (Verlag für Orientkunde Dr. H. Vorndran) 1968 (Beiträge zur Sprach- und Kulturgeschichte des Orients 19).

ThoṬab D. Thomas, "al-Ṭabarī", in: *EI*[2] [q.v.] 10/17a–18b.

TīfAz Aḥmad ibn Yūsuf at-Tīfāšī, *Azhār al-afkār fī ǧawāhir al-aḥǧār*, ed. M.Y. Ḥasan and M. Basyūnī, al-Qāhira (al-Hai'a al-Miṣrīya al-'Āmma lil-Kitāb) 1397/1977.

UllIM Manfred Ullmann, *Islamic Medicine*, Edinburgh (Edinburgh University Press) 1978 (Islamic Surveys 11).

UllMed Manfred Ullmann, *Die Medizin im Islam*, Leiden–Köln (E.J. Brill) 1970 (Handbuch der Orientalistik, Erste Abteilung, Ergänzungsband 6,1).

UllNGw Manfred Ullmann, *Die Natur- und Geheimwissenschaften im Islam*, Leiden (E.J. Brill) 1972 (Handbuch der Orientalistik, Erste Abteilung, Ergänzungsband 6,2).

VāgAṣṭ *Vāgbhaṭa's Aṣṭāñga Hṛdayam*, 1–3, ed. and tr. K.R. Srikantha Murthy, Varanasi (Chowkhamba Krishnadas Academy) 2010 (reprint) (Krishnadas Ayurveda Series 27).

VatKam *Vatsyayana Mallanaga Kamasutra*, tr. Wendy Doniger and Sudhir Kakar, Oxford (Oxford University Press) 2003 (Oxford World's Classics).

VuLex Johann August Vullers, *Lexicon Persico-Latinum Etymologicum*, 1–2, Graz (Akademische Druck- & Verlagsanstalt) 1962 (reprint of the edition Bonn 1855–1864).

WaṬab Elvira Wakelnig, "Al-Ṭabarī and al-Ṭabarī. Compendia between Medicine and Philosophy", in: *Philosophy and Medicine in the Formative Period of Islam*, ed. Peter Adamson and Peter E. Pormann, London (The Warburg Institute) 2017 (Warburg Institute Colloquia 31).

WatStu Gilbert Watson, *Theriac and Mithridatium. A Study in Therapeutics*, London (The Wellcome Historical Medical Library) 1966 (Publications of the Wellcome Historical Medical Library, New Series 9).

WeiZit Ursula Weisser, "Die Zitate aus Galens De Methodo Medendi im Ḥāwī des Rāzī", in: *The Ancient Tradition in Christian and Islamic Hellenism. Studies on the Transmission of Greek Philosophy and Sciences Dedicated to H.J. Drossaart Lulofs on his Ninetieth Birthday*, ed. Gerhard Endress and Remke Kruk, Leiden (Research School CNWS) 1997 (CNWS Publications 50), pp. 279–318.

WkaS Manfred Ullmann, *Wörterbuch der klassischen arabischen Sprache*, 1–2 [in 5 vols], Wiesbaden (Otto Harrassowitz) 1970–2009.

WüAN Ferdinand Wüstenfeld, *Geschichte der arabischen Aerzte und Naturforscher, nach den Quellen bearbeitet*, Göttingen (Vandenhoeck & Ruprecht) 1840.

YāBul Yāqūt ibn ʿAbdallāh al-Ḥamawī, *Jacut's geographisches Wörterbuch* [*Muʿǧam al-buldān*], 1–6, ed. Ferdinand Wüstenfeld, Leipzig (F.A. Brockhaus) 1866–1870.

YāIr Yāqūt ibn ʿAbdallāh al-Ḥamawī, *Dictionary of Learned Men* [*Iršād al-arīb*], 1–7, ed. D.S. Margoliouth, Leyden–London (E.J. Brill and Luzac & Co.) 1907–1927 (E.J.W. Gibb Memorial Series 6,1–7).

ZauMuḫ Muḥammad ibn ʿAlī al-Ḫaṭībī az-Zauzanī, *Taʾrīḫ al-ḥukamāʾ wa-huwa muḫtaṣar az-Zauzanī al-musammā bil-Muntaḫabāt al-multaqaṭāt min kitāb Iḫbār al-ʿulamāʾ bi-aḫbār al-ḥukamāʾ li-Ǧamāladdīn Abī l-Ḥasan ʿAlī ibn Yūsuf al-Qifṭī*, ed. Julius Lippert auf Grund der Vorarbeiten Aug[ust] Müller's, Leipzig (Dieterich'sche Verlagsbuchhandlung) 1903.

Indices

Introduction to the Indices

All numbers in the following indices refer to paragraphs (not pages) which run consecutively throughout the edition of the Arabic text and its English translation (pp. 38–161); occasionally, a number may be preceded by an asterisk (∗), in which case the reference is to a footnote in the English translation. The first and most important inventory (medicine and pharmacy) covers words and terms relating to substances and products, pathology and anatomy, medico-pharmaceutical implements, therapeutic procedures, applicative categories, generics, as well as the environment, health workers, patient groups, and iatromathematics—here, for reasons of economy, pure verbal constructions in the base text and/or non-substantive renditions on my part have, with few exceptions, not been registered; for the same reason, transliterated Arabic lemmata (and their English equivalents) are given, again with some exceptions, as nomina singularis or collectiva, regardless of the particular morphological and grammatical form they may assume in the base text; primary or elementary qualities of substances are filed under the respective nominal forms. The Arabic–English sub-inventory (pp. 189–201) follows the sequence of the Roman alphabet, whose order is not influenced by diacritics. The abbreviation 'ssc' before a number refers to the chapter-heading (*superscriptio*) that precedes the respective paragraph. Chevrons in the taxonomic inventory (p. 206 f.) contain cross-referential aids. If these indices are more comprehensive and more detailed than appears to be warranted by the relatively small size of the underlying text, the user is asked to remember that Ṭabarī wrote at a time when the Arabic language went through a crucial phase of terminological and conceptual formation well worth documenting; aside from this general observation, there has also been, on my part, the wish to establish a thorough lexicographical record of Ṭabarī's particular style of writing and targeted narrative choices.

Index of Medicine and Pharmacy

a English–Arabic

b Arabic–English

ʿain → eye; spring
ʿain aš-šams → eye of sun
ʿair → onager
aiyām as-sana → time of year
aiyil → stag
akalāt → food
ākila → sore, gangrenous
akl → food
akla → meal
ʿalaf → forage
ʿalāma → sign; symptom
ʿamal → toil; work
ʿāmil → yoke-ox
amlaǧ → emblic
ʿāna → region, pubic
anbaǧa → preserve
ʿanbar → ambergris
anf → nose
ānisa → lover
anwāʿ al-ʿilal → conditions, pathological
ʿanzarūt → sarcocolla
ʿāqiba → consequence
ʿaql → mind
ʿaqqār → drug
ʿaqrab → Scorpio; scorpion
ʿaraq → perspiration; sweat
arḍ → earth; ground; soil
arḍ qāʿ ḥarra → land, flat and stony
arḍ šaǧrāʾ → land, bosky
arnab → hare
aruzz → rice
ās → myrtle
ʿašāʾ → evening meal
ʿaṣab → nerve
asad → Leo
aṣal → habit, bad
ʿasal → honey
ʿašan → nightblindness
ašfiya wa-manāfiʿ → applications, therapeutic
aṣḥāb at-taʿab wal-kadd → labourers, heavy
aṣiḥḥāʾ → people, healthy
asqāʾ → lands, irrigated
ašyāʾ ʿalā infirādihā → substances, simple
ašyāʾ min al-ḥayawānāt → meats
ašyāʾ min an-nabt → vegetables
atān → donkey, female
ʿaṭaš → thirst
aṯfāl → dregs
aṭʿima wa-aġḏiya → foodstuffs
attūn → kiln
ʿaura → genitals
ʿausaǧ → boxthorn
āya → sign
azm → restraint
ʿaẓm → bone
ʿaẓm al-insān → bone, human

bābūnaǧ → chamomile
badan → body
bāḏarūǧ → basil
bāḏinǧān → aubergine
bāḏirnaǧbūyah → lemon balm
baġla → mule, female
bāh → coitus; intercourse, sexual; sex
bahāʾim → beasts; cattle
bahaq → vitiligo
baḥr → lake; sea water
baiḍ → eggs
baiḍ nīmbirišt → eggs, poached
bait → chamber; house
balad → land
balāda → dullness; stupidity
balāḏur → marking-nut
balaḥ → date, unripe
balġam → phlegm
balīlaǧ → myrobalan, beleric
banafsaǧ → sweet violet
baqāʾ → existence; life; subsistence; survival
baqar → cattle; cows
bāqillā → broad bean
baql → plant
baqla yamānīya → blite
baʿr → dung
baraš/ṣ → leprosy, white
bard → cold(ness); coolness
barīya → plain
barsām → phrenitis; pleurisy
bārzad → galbanum
bašāʿa → foulness
baṣal → onion
baṣar → sight
bašara → skin
basbāsa → mace
baṭāʾa → laziness
baṭīḥa → valley bottom
baṭn → belly
baṭš → willpower
baṭṭ → duck

ǧušāʾ → belch(ing)
ġusl → washing
ǧuzʾ → organ
ǧuzair → baby carrot

ḥabaq → basil
ḥabaq nahrī → watermint
ḥabb → pill
ḥabba ḫaḍrāʾ → pistachio
ḥabīb → friend; lover
ḫabz → baking
ḥadaqa → pupil
ḫadar → numbness
haḏayān → hallucination
ḥadd → boundary
ḥadīd → iron
haḍm → digestion
ḫafaqān → palpitations
ḥāfir → hoof
ḥaǧal → partridge
ḥaǧar al-maġnāṭīs → lodestone
ḥāǧat kull insān → need, individual
ḥāǧāt → times of need
ḥāǧib → brow
haiḍa → diarrhoea, vomitive
ḥaiya → snake
hāl → cardamom
ḫalāʾ → depletion
halāk → demise
ḥalāwa → sweetness; taste, sweet
ḥalīla → wife
halīlaǧ → myrobalan
ḥall → sesame oil
ḫall → vinegar
ḫall ḫamr → wine vinegar
ḥalq → throat
ḥalwā → pastry
ḥamal → Aries
ḥamām → pigeons
ḫamīr → dough, leavened; leaven
hamm → sorrow; worrying
hāmma → creeper; vermin
ḥammām → bath(house)
ḥammāra wa-ḥarāra → furnace
ḫamr → wine
ḥanak → palate
ḫanāzīr → scrofula
ḥandaqūqā → sweet clover
ḫaniṯa → woman, sensual
ḥanẓal → colocynth
ḥarāfa → acridness
ḥaraka → exercise; movement
haram → ageing
haram ʿaraḍī → decline, accidental
haram ṭabīʿī → decline, natural
ḥarāra → heat; temperature; warmth
ḥarāra (ġarīzīya) → heat, innate
ḥarārāt → fever heat
ḥarb → war
ḫardal → mustard
ḫarīf → autumn
harim → man, old
ḥarīra → cloth of silk
ḥarmal → Syrian rue
harnuwa → aloe berry
ḥarq (an-nār) → burn
ḥarr → heat; warmth
ḫarūf → lamb
ḫašaba → stick
ḥasad → envy
ḥaṣā(t) al-maṯāna → bladder stones
ḫašḫāš abyaḍ → poppy, white
ḫašḫāš aswad → poppy, black
ḫaṣīy → eunuch
ḥašīya → cushion
ḫass → lettuce
ḫāṣṣīya → property, special
ḥasw → broth
ḥaṭab ǧazl → wood chunks, thick
ḫaṭmī → marshmallow
ḫauf → fear
ḫauḫ → peach
ḥaulī ḍ-ḍaʾn → sheep yearling
hawāʾ → air; temperature
hawaǧ → foolishness
ḥawāʾiǧ → duties, daily
ḫawar → fragility
ḥawāss → senses
hāwun → mortar
hayaǧān → agitation; eruption
ḥayāh → life
ḥayawān → animal; creature; living being
ḫazaf → jar of clay
ḥazāz → scurf
ḥibr → ink
ḫiḍāb muǧarrab → dye, well-tried
ḥidda → abruptness; irascibility
ḫiffa → frivolity; lightness

waǧh → face
wahn → debility
waʿl → mountain goat
walad → child
waqt al-ġiḏāʾ → mealtime
waram → swelling; tumour
warašān → ringdove
ward → rose
wark → hip
wasaḫ → earwax; filth
waswasa → hearing voices
wilāda → childbirth
wisād → bed
wuḍūʾ → ablution

yabrūḥ → mandrake
yad → hand
yamīn → side, right
yaraqān → jaundice
yāsamīn → jasmine
yasār → side, left
yubs → desiccation; dryness; staleness
yubūsa → dryness

zabīb → raisin
zaʿfarān → saffron
zaḥīr → dysentery
ẓahr → back
zait → olive oil
zait rikābī → Syrian olive oil
zaitūn → olive tree
zamān → season
zamān (*as-sana*) → time of year
zanbaq → jasmine oil
zanǧabīl → ginger
zanǧabīl ṣīnī → Chinese ginger
zarʿ → semen
zaʿtar → savory
zauǧ → spouse
zawaǧān → dishonesty
zibl → droppings; dung
zibl al-insān → feces, human
zifira → greasiness
zīrbāǧa → spoon-meat
ẓufr → fingernail
zuǧāǧ → jar of glass
zuḥal → Saturn
zuhara → Venus
zukām → catarrh
ẓulma → dark(ness)
zunbūr → hornet
zurzūr → starling

Index of People and Places

Index of Work Titles

Index of Miscellaneous Terms

Index of Botanical Names

Appendix

The textual relationship between Ṭabarī's *Paradise of Wisdom* and his *Health Regimen* can be illustrated through the following chart:*

Paradise of Wisdom	→	***Health Regimen***
p. 4,5 f.	∽	§ 2 *fa-lā sabīl … al-arbaʿ*
pp. 129,4–8; 21 ff.; 522,11 ff.	≈	§ 8 *wa-dafʿ … ar-rāḥa*
pp. 99,21–100,2	∽	§ 10 *inna ḥifẓ … ar-radīʾa*
pp. 99,18–21; 102,11–18	≈	§ 11 *innahū lammā … li-ḍīq manāfiḏihī*
p. 118,2 ff.	∽	§ 12 *inna l-waǧh … al-bārida al-muʿtadila*
pp. 103,19 ff.; 569,18–21	≈	§ 12 *wa-iḏā kānat al-maʿida … diqāq al-ḥaṭab*
p. 103,24 f.	≈	§ 13 *fa-ammā d-dalīl … al-ʿain*
p. 112,11 ff.	∽	§ 13 *wa-mā yuʿīn … al-istiyāk*
p. 100,4–10	∽	§ 14 *yanbaġī … yusaḫḫinuhū*
pp. 117,9 ff.; 22–118,1	≈	§ 14 *wa-yabdaʾ … bi-fasādihā*
pp. 100,2 f.; 11 ff.; 101,14 f.	≈	§ 15 *inna mimmā … murtafaʿan*
pp. 100,22–101,1; 110,6 ff.	≈	§ 15 *wa-in ʿaṭiša … māliḥa*
p. 101,1 f.	∽	§ 16 *inna … al-kabid*
p. 100,13 ff.	∽	§ 17 *inna … asqāman*
p. 100,15–18	∽	§ 18 *an-naum qabla … al-inḍāǧ*
p. 101,3–12	≈	§ 19 *inna man … qaʿrihī*
p. 111,16–19	≈	§ 21 *fa-ammā mā yusamminuhū … ar-radīʾa*
p. 111,19 f.	≈	§ 21 *wa-anfaʿ … ar-raǧāʾ*
pp. 111,20–112,1	≈	§ 21 *fa-ammā mā yuhazzil … al-yābisa*
p. 112,1 f.	≈	§ 21 *wa-aqwā … al-humūm*
pp. 112,21–25; 120,7 ff.	≈	§ 22 *wa-mā yufarriḥ … tanfuḥ*
p. 116,17–20	≈	§ 24 *inna l-maṭʿam … ad-dāʾ*
p. 87,19 f.	∽	§ 24 *kas-sirāǧ … inṭafaʾa*
pp. 41,13–16; 208,11–17	≈	§ 27 *fa-innahā … lahā*

* In this chart, the sign ∽ indicates closer correspondences, the sign ≈ points to looser parallels; in a few (exceptional) cases, the sign = has been used to highlight literal borrowings. I was able to consult here, with some benefit, the verifications already made by Usama Raslan in the footnotes to his German translation of the *Health Regimen*, cf. ṬabHT 21–127 *passim*.

 | DOI:10.1163/9789004445895_006

(*cont.*)

Paradise of Wisdom	→	*Health Regimen*
pp. 5,1 ff.; 5–8; 133,3 f.; 339,19 ff.	≈	§ 29 *wa-li-hāḏā ... al-miʿā*
pp. 88,22–89,6	≈	§ 30 *wa-ka-ḏālika ... fa-taṭfaʾ*
pp. 105,22 f.; 106,2–5; 8 ff.	≈	§ 33 *inna r-rabīʿ ... al-ḥammām*
pp. 106,19–23; 107,3 ff.; 11 ff.	≈	§ 34 *fa-ammā ṣ-ṣaif ... al-ḥāǧa ilaihī*
pp. 107,20 f.; 108,1; 3 ff.; 7 f.; 10 f.	≈	§ 35 *al-ḫarīf ... ad-dam fīhī*
p. 108,15 f.	∽	§ 36 *inna fuḍūl ... yuẓhiruhā*
p. 108,17–20	∽	§ 37 *inna l-buṭūn ... yuʾkal fī l-ḫarīf*
pp. 108,20–23; 109,2 f.; 7 f.	≈	§ 38 *wa-an yuʾkal ... yusaḫḫanuhū*
p. 340,8–13	≈	§ 39 *inna fī ... ruṭūbatihī*
pp. 340,19–341,5; 8 f.; 16 ff.; 20 f.	≈	§ 40 *wa-min ṣawāb ... bil-qusṭ wal-qaranful*
p. 104,1–7	∽	§ 41 *inna l-badan ... auǧāʿ ar-raʾs*
p. 104,8–12	∽	§ 42 *wal-ǧuzʾ ... al-ḥummā*
p. 104,13–16	∽	§ 43 *wal-ǧuzʾ ... al-bawāsīr*
p. 104,17–23	∽	§ 44 *wal-ǧuzʾ ... ar-rabw*
pp. 104,24–105,11	∽	§ 45 *wa-ḏakarat ... ar-ribʿ*
pp. 114,21 f.; 115,3	∽	§ 46 *inna l-insān ... al-baqāʾ*
pp. 118,10 f.; 13 ff.; 119,2 ff.; 8–12	≈	§ 48 *inna min ... ašbahahā*
p. 119,12–15; 19 f.	≈	§ 49 *wa-innamā ... aṭ-ṭaʿm*
p. 384,11–21	≈	§ 50 *wa-mā kāna ... al-ḫall waṯ-ṯūm*
pp. 388,17–21; 389,2 f.	≈	§ 52 *wa-kāna fī miyāh ... fī l-bāh*
p. 384,16–20	≈	§ 53 *wa-kull dābba ... al-ǧabalīyāt*
p. 386,12–18	≈	§ 54 *wa-taḫtalif ... al-utun*
p. 391,6–9; 15 f.	≈	§ 55 *fa-mā kāna ... laṭāfa*
p. 391,5 f.	≈	§ 55 *wa-kullamā ... ġiḏāʾan*
pp. 389,12; 390,1; 3 f.; 14 f.	≈	§ 56 *wa-kull duhn ... bainahumā*
p. 394,5	∽	§ 56 *wa-kull kāmaḫ ... minhū*
p. 393,16–20	≈	§ 57 *wal-ḫall ... ġaiyarahumā*
p. 392,6 ff.; 10 ff.; 20 f.	≈	§ 58 *wa-qūwa ... yuqauwīhā wa-yaḥbis al-baṭn*
p. 11,18 ff.	∽	§ 62 *faṭ-ṭabāʾiʿ ... wa-humā r-ruṭūba wal-yubūsa*
p. 20,14 ff.; 20 f.	≈	§ 62 *wa-qawām ... al-ḥayawān*
pp. 42,15 ff.; 85,3 ff.; 124,23–125,1	≈	§ 63 *fa-ammā d-dam ... al-ḥārra*
p. 40,12–15	≈	§ 63 *wa-yakūn tawallud ... as-saudāʾ*
pp. 42,12 ff.; 85,5 f.; 125,1 f.	≈	§ 64 *fa-ammā l-mirra ... fī ṣ-ṣafrāʾ*
pp. 42,19 f.; 85,7 f.; 125,2 ff.	≈	§ 65 *wa-ammā l-balġam ... aṣ-ṣarʿ*

(*cont.*)

Paradise of Wisdom	→	*Health Regimen*
pp. 42,22–43,1; 85,6 f.; 125,4 f.	≈	§ 66 *wa-ammā s-saudāʾ … as-saraṭān*
p. 58,11 f.	∽	§ 68 *wa-hāḏā … al-afʿāl*
p. 59	∽	§ 68 *Diagram*
p. 125,9–19	≈	§ 69 *yakūn … al-baṭn*
p. 125,19–126,3	≈	§ 70 *wa-yakūn … ḥarāratuhū*
p. 126,6–9	≈	§ 71 *wa-yakūn … ar-radīʾa*
p. 126,3–6	≈	§ 72 *wa-yakūn … aṭ-ṭaʿām*
p. 45,14 f.; 17 f.	≈	§ 74 *inna man … at-tarkīb*
pp. 49,14–50,2	≈	§ 74 *wa-in kānat ṣūratuhū … ǧimāʿ*
pp. 93,16 f.; 126,16 ff.	≈	§ 75 *fa-min ad-dalīl … al-ḥasana*
pp. 93,17 f.; 126,15 f.	≈	§ 76 *wa-min ad-dalīl … aš-šaġab*
pp. 93,19 f.; 126,22 ff.	≈	§ 77 *wa-min ad-dalīl … ḏālika*
pp. 93,18 f.; 126,20 ff.	≈	§ 78 *wa-min ad-dalīl … at-tahāwīl*
p. 565,3 ff.	=	§ 80 *inna man … ḥāǧātihī*
p. 565,6–13	=	§ 81 *wa-yastāk min … al-ḫafaqān*
p. 565,14–18	=	§ 82 *fa-iḏā … waram*
p. 565,18–22	=	§ 83 *ṯumma … al-ḥāmil*
pp. 565,22–566,1	=	§ 84 *ṯumma … al-bāh*
p. 566,1 ff.	=	§ 85 *ṯumma … ḫumār*
p. 566,4–7	=	§ 86 *fa-iḏā … ʿalā ḏālika kulluhū*
p. 566,7–12	=	§ 87 *fa-iḏā … as-suʿāl*
p. 566,12–19	=	§ 88 *fa-iḏā … al-qaiʾ*
pp. 566,19–567,1	=	§ 89 *fa-iḏā … aṭ-ṭīb*
p. 567,1 ff.	=	§ 90 *ṯumma … ṣadrihī*
p. 570,13 f.; 17–21	∽	§ 90 *wa-yabdaʾ … baṭīʾ^{an}*
p. 567,7–19	≈	§ 91 *wal-yaḥḏar … az-zarʿ*
pp. 567,23 f.; 568,2 f.	∽	§ 92 *inna l-ikṯār … al-haram*
p. 581,2–9	∽	§ 93 *inna ʿilal … al-ḥasad*
p. 568,13–23	∽	§ 94 *inna l-māʾ … tilka l-arḍ*
p. 569,2–6	∽	§ 94 *wa-innamā … ḥaiyāt*
p. 569,11 ff.	∽	§ 94 *inna šarb … al-balġam*
p. 571,19	=	§ 95 *fa-ammā … al-āḫira*
p. 572,1 f.	∽	§ 95 *wa-ʿaǧab^{an} … yamraḍ*
p. 572,19–24	≈	§ 95 *fa-inna š-šarāb … al-bārida*
p. 573,1 f.	≈	§ 95 *wa-man arāda … yadaʿuhā*
p. 569,18–21	≈	§ 96 *inna l-aṭʿima … lahabuhā*

(*cont.*)

Paradise of Wisdom	→	*Health Regimen*
p. 570,8	=	§ 96 *wa-qad yakūn ... maraḍ*
p. 399,14–19	∽	§ 97 *inna kull ... lil-ḫanāfis*
p. 356,7–19	∽	§ 98 *wa-li-kull ... aḏ-ḏiʾb*
pp. 358,5 ff.; 360,6 f.	∽	§ 99 *inna l-maḏāqāt ... ḏākiruhā*
pp. 358,7 ff.; 18 f.; 359,3 ff.; 8 ff.; 360, 11; 22 ff.; 361,1 f.; 362,1–4	≈	§ 100 *fa-aqrabuhā ... bil-bard wal-yubs*
p. 374,12–17	≈	§ 102 *fa-auwaluhā ... anḍağahū*
p. 375,1 ff.	∽	§ 103 *aš-šaʿīr ... aṣ-ṣadr*
p. 375,5 ff.	∽	§ 104 *al-aruzz ... bil-laban*
p. 375,9	∽	§ 105 *al-ğāwars ... al-baṭn*
p. 375,10 f.	∽	§ 106 *al-bāqillā ... liṣ-ṣadr*
p. 375,12–15	∽	§ 107 *at-turmus ... al-baṭn*
p. 375,15 f.	∽	§ 108 *al-ḥimmaṣ ... al-kabid*
p. 376,7 f.	∽	§ 109 *al-lūbiyāʾ ... ġilaẓ^{an}*
p. 376,8 f.	∽	§ 110 *al-ʿadas ... ad-dam*
p. 376,9 f.	∽	§ 111 *al-ḥulba ... aṣ-ṣadr*
p. 376,13	=	§ 112 *as-simsim ... al-manīy*
p. 376,13 f.	∽	§ 113 *al-ḫašḫāš ... fīhī*
p. 376,17	∽	§ 114 *al-qurṭum ... al-baṭn*
p. 376,16	∽	§ 115 *aš-šahdānağ ... al-manīy*
p. 377,8 f.	∽	§ 117 *auwal ... yunauwim*
p. 377,14–18	≈	§ 118 *al-hindabāʾ ... al-yaraqān*
p. 377,21 f.	=	§ 119 *al-ğirğīr ... aṣ-ṣudāʿ*
pp. 377,23–378,2	∽	§ 120 *al-fuğl ... al-yaraqān*
p. 378,2 f.	∽	§ 121 *as-salğam ... al-manīy*
p. 380,20 f.	∽	§ 122 *al-ğazar ... yunḍiğ*
p. 378,3 f.	∽	§ 123 *al-karafs ... al-maʿida*
p. 378,4 ff.	=	§ 124 *an-naʿnaʿ ... al-fuwāq*
p. 378,6 f.	∽	§ 125 *al-ḥabaq ... al-hawāmm*
p. 378,8 f.	∽	§ 126 *al-kuzbara ... ḫadar^{an}*
p. 378,9 f.	∽	§ 127 *al-hilyaun ... al-manīy*
p. 378,10 f.	∽	§ 128 *al-bāḏirnağbūyah ... al-qalb*
p. 378,11	∽	§ 129 *aṭ-ṭarḫūn ... ṯaqīl*
p. 378,14 ff.	∽	§ 131 *al-kurrāṯ ... miṯluhū*
p. 379,1–4	∽	§ 132 *aṯ-ṯūm ... al-qarawīyīn*
p. 378,16–20	≈	§ 133 *al-baṣal ... al-miyāh*

(*cont.*)

Paradise of Wisdom	→	*Health Regimen*
p. 379,6 f.	∽	§ 134 *al-bāḏinǧān … as-saudā'*
p. 379,9 ff.	∽	§ 135 *al-kurunb … aš-šarb*
p. 379,12 ff.	∽	§ 137 *as-silq … bil-ma'ida*
p. 379,14 f.	∽	§ 138 *al-lablāb … aṣ-ṣafrā'*
p. 379,16–20	∽	§ 139 *al-kabar … ad-dūd*
p. 380,14 ff.	≈	§ 140 *al-kašūṯ … al-kabid*
p. 379,21	≈	§ 141 *as-sarmaq … al-qai'*
p. 379,22 f.	∽	§ 142 *al-baqla … al-ḥarr*
p. 379,23	=	§ 143 *al-isfānāḫ … al-yamānīya*
pp. 379,24–380,1 f.	∽	§ 144 *al-bāḏarūǧ … ar-ru'āf*
p. 380,3 f.	∽	§ 145 *aṣ-ṣa'tar … al-ġalīẓa*
p. 380,4 f.	∽	§ 146 *aš-šibiṯṯ … al-qai'*
p. 380,6 f.	∽	§ 147 *as-saḏāb … lil-ǧimā'*
p. 380,8 f.	=	§ 148 *al-ḫardal … yubsihī*
p. 380,10 f.	=	§ 149 *al-ḥurf … al-ġalīẓa*
p. 380,23 ff.	≈	§ 150 *al-ḥandaqūqā … al-balġam*
pp. 380,25–381,1	∽	§ 151 *ar-rāsan … al-bard*
p. 381,3 f.	∽	§ 152 *al-ḥummāḍ … al-am'ā'*
p. 381,5 f.	∽	§ 153 *'inab … sadadihimā*
p. 394,5	∽	§ 156 *wa-kull … aš-šai'*
p. 381,13 ff.	∽	§ 157 *'āmma … al-baṭn*
p. 381,15–19	∽	§ 158 *wa-ra's … al-aswad*
p. 381,20 f.	∽	§ 159 *wal-'inab … qalīl^an^*
pp. 381,22–382,1	∽	§ 160 *az-zabīb … al-baṭn*
p. 382,2 f.	∽	§ 161 *al-firṣād … yaḍurr*
p. 382,4 f.	∽	§ 162 *ḥabb … al-baṭn*
p. 382,5 ff.	∽	§ 163 *at-tuffāḥ … bil-'aṣab*
p. 382,11 ff.	∽	§ 164 *al-utruǧǧ … duhn*
p. 382,18	∽	§ 165 *al-iǧǧāṣ … yuḫriǧuhā*
p. 382,19	≈	§ 166 *at-tamr … mulaiyin*
p. 382,19	∽	§ 167 *an-nabiq … qabbāḍ*
p. 382,20	∽	§ 168 *as-safarǧal … yadbuġuhā*
p. 382,22 f.	∽	§ 169 *ar-rummān … al-baṭn*
p. 383,3 f.	∽	§ 170 *ṯamar … qabbāḍ*
p. 383,4	∽	§ 171 *al-mauz … lil-ma'ida*
p. 383,2 f.	∽	§ 172 *al-kummaṯrā … al-baṭn*

(*cont.*)

Paradise of Wisdom	→	*Health Regimen*
p. 383,5 f.	∽	§ 173 *at-tamr … min al-balaḥ*
p. 383,7 f.	∽	§ 174 *al-ʿunnāb … ar-rabw*
p. 383,9	∽	§ 175 *al-ġubairāʾ … al-baṭn*
p. 383,9 f.	∽	§ 176 *al-ǧauz … sarīʿ^an^*
p. 383,11 ff.	∽	§ 177 *al-ǧillauz … al-ʿaqārib*
p. 383,15–18	∽	§ 178 *al-lauz … al-kabid*
p. 383,18 f.	∽	§ 179 *ḥabb … lil-bāh*
p. 383,24 f.	∽	§ 180 *al-ḥabba … al-fāliǧ*
p. 384,3 f.	∽	§ 181 *ǧauz … al-manīy*
p. 384,5 f.	∽	§ 182 *al-ḫauḫ … al-ḫilṭ*
p. 379,9	≈	§ 183 *al-kamʾa … radīʾ*
p. 393,3 ff.	≈	§ 207 *al-halīlaǧ … dūnahū*
p. 393,9 f.	≈	§ 209 *az-zanǧabīl … al-burūdāt*
p. 393,9 f.	≈	§ 210 *aš-šašqāqul … al-māʾ*
p. 393,7 f.	≈	§ 211 *al-qarʿ … yasīr^an^*
p. 568,13	∽	§ 215 *al-māʾ … nabt*
p. 504,5–9	∽	§ 215 *wa-afḍal … ġilaẓihī*
p. 504,14 f.	∽	§ 215 *wal-miyāh … ṯaqīla*
pp. 504,22–505,1	∽	§ 215 *wa-miyāh al-baṭāʾiḥ … al-kabid*
p. 505,6 f.; 12–19	≈	§ 215 *wa-miyāh al-ʿuyūn … ʿuḍw bārid*
p. 420,4 f.; 7; 11–14; 19 f.	∽	§ 216 *inna šaʿr … baiyin^an^*
p. 421,6–10; 17 f.	∽	§ 217 *wa-in uḫiḏa … al-wilāda*
pp. 421,21; 422,2 f.; 6–10; 14 f.; 20–25	≈	§ 218 *wa-in ṣubba … as-simām*
pp. 423,16 f.; 424,3 ff.	∽	§ 219 *wa-in uḫiḏat … az-zaḥīr*
p. 424,12 f.; 15 f.	∽	§ 220 *wa-luḥūm … al-waǧaʿ*
p. 425,4–7; 18–21	∽	§ 221 *wa-in šuwiya … al-mafāṣil*
p. 426,15 f.; 18 ff.	∽	§ 222 *baul … yaḍurruhū l-kalb al-kalib*
p. 427,7 ff.	∽	§ 224 *wa-laban … ar-ruʿāf*
p. 429,3 f.	∽	§ 225 *wa-in uḫiḏat … ǧilduhū*
p. 430,6 ff.	∽	§ 226 *wa-in uḫiḏa … al-yusrā*
p. 430,11–14	∽	§ 227 *in wuḍiʿat … aš-šauk*
p. 431,3; 9–13	≈	§ 228 *wa-in duhina … al-fazaʿ*
pp. 431,19 ff.; 432,1 f.	≈	§ 229 *wa-in uḫiḏa … as-sill*
pp. 432,5–8; 14 f.; 433,2 f.	∽	§ 230 *wa-in waḍaʿta … al-qaulanǧ*
p. 433,13 f.	≈	§ 231 *wa-in ḫuliṭa … ǧalāhū*
pp. 433,21 f.; 434,5 ff.	≈	§ 232 *fa-ammā l-ḥamām … al-maṯāna*

(*cont.*)

Paradise of Wisdom	→	*Health Regimen*
p. 435,3	∽	§ 233 *wa-in uḫiḏa … aṣ-ṣarʿ*
p. 435,4	≈	§ 234 *wa-qad tanfaʿ … minhā*
p. 435,9 f.	≈	§ 235 *wa-luḥūm … nafaʿahā*
p. 437,19 f.	∽	§ 236 *wa-in uḥriqa … aš-šaʿr*
p. 437,22 ff.	≈	§ 237 *wa-in uḫiḏa … nafaʿa*
p. 438,4–7	≈	§ 238 *wal-ǧundbādastar … al-ḫafaqān*
p. 438,14 ff.; 19	≈	§ 239 *wa-in uḫiḏa … al-ḥaiyāt*
p. 440,5–8; 13 f.	∽	§ 240 *wa-in uḥriqa … ḏālika*
p. 441,7 ff.; 11 f.; 15 ff.	≈	§ 241 *wa-in uḫiḏa … al-baṣar*
p. 441,23 f.	≈	§ 242 *al-ʿaqrab … al-maṯāna*
p. 442,2–5	∽	§ 243 *wa-sāmm … minhū*
p. 410,15 f.	∽	§ 244 *al-midād … baiyin*[an]
pp. 410,23–411,1	≈	§ 245 *wa-ka-ḏālika … baiyin*[an]
pp. 411,24–412,1	≈	§ 246 *aṭ-ṭīn … al-armanī*
pp. 135,24–136,4	≈	§ 251 *dawāʾ … al-ḫaṭmī*
pp. 451,24–452,2	≈	§ 252 *ṣifa … bi-duhn al-ḥall*
pp. 467,21–468,2	∽	§ 254 *ṣifa … fātir*
p. 166,2 ff.	≈	§ 255 *dawāʾ … bihī*
p. 145,19–22	≈	§ 257 *dawāʾ … marrāt*
pp. 492,20–493,7	≈	§ 258 *dawāʾ … al-marʾa zina dirhamain*
p. 447,12 ff.	≈	§ 258 *wa-in uḫiḏa … nafaʿahū*
pp. 446,18 ff.; 24–447,11	∽	§ 259 *dawāʾ … fa-huwa afḍal*
p. 446,21 ff.	≈	§ 259 *wa-yusqā … ilā l-māʾ*
p. 446,6	∽	§ 259 *wa-minhum … maṯal al-kilāb*
p. 602	≈	§ 261 *Diagram*

Printed in the United States
By Bookmasters